# Staging Family Science Nights

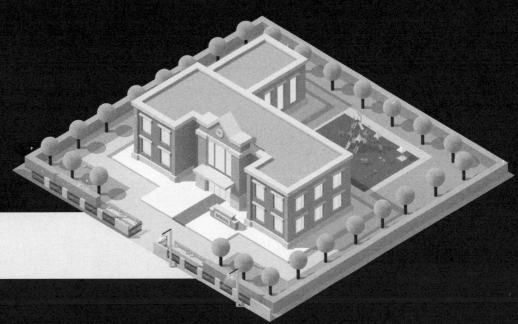

# Staging Family Science Nights

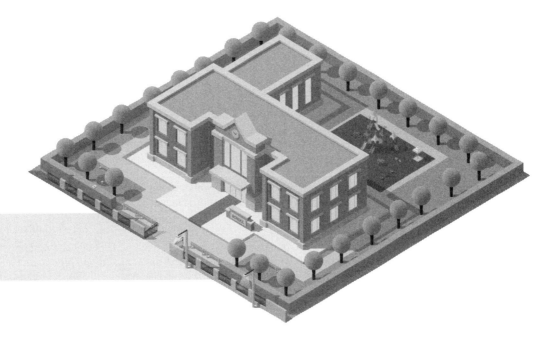

Donna Governor and Denise Webb

National Science Teachers Association

Arlington, VA

National Science Teachers Association

Claire Reinburg, Director
Rachel Ledbetter, Managing Editor
Andrea Silen, Associate Editor
Jennifer Thompson, Associate Editor
Donna Yudkin, Book Acquisitions Manager

ART AND DESIGN
Will Thomas Jr., Director
Joe Butera, Senior Graphic Designer, cover design
Capitol Communications LLC, interior design

PRINTING AND PRODUCTION
Catherine Lorrain, Director

NATIONAL SCIENCE TEACHERS ASSOCIATION
David L. Evans, Executive Director

1840 Wilson Blvd., Arlington, VA 22201
*www.nsta.org/store*
For customer service inquiries, please call 800-277-5300.

*NSTA is committed to publishing material that promotes the best in inquiry-based science education. However, conditions of actual use may vary, and the safety procedures and practices described in this book are intended to serve only as a guide. Additional precautionary measures may be required. NSTA and the authors do not warrant or represent that the procedures and practices in this book meet any safety code or standard of federal, state, or local regulations. NSTA and the authors disclaim any liability for personal injury or damage to property arising out of or relating to the use of this book, including any of the recommendations, instructions, or materials contained therein.*

**Library of Congress Cataloging-in-Publication Data**
Names: Governor, Donna, 1957- author. | Webb, Denise, 1968- author.
Title: Staging family science nights / by Donna Governor and Denise Webb.
Other titles: Family science nights
Description: Arlington, VA : National Science Teachers Association, [2018].
Identifiers: LCCN 2018039313 (print) | LCCN 2018051160 (ebook) | ISBN 9781681406244 (e-book) |
    ISBN 9781681406237 (print)
Subjects: LCSH: Science--Study and teaching (Elementary)--Activity programs. | Science--Study and
    teaching--Activity programs. | Science--Study and teaching--Parent participation. | Student-centered
    learning. | Project method in teaching. | Science and the arts. | Education in the theater. | Education--
    Parent participation. | Community theater. | Family recreation.
Classification: LCC Q164 (ebook) | LCC Q164 .G66 2018 (print) | DDC 372.35--dc23
LC record available at *https://lccn.loc.gov/2018039313*

# Table of Contents

# Preface

## It Takes a Community

There's an ancient tradition in the Western world called *community theater*. In classical Greece, it was common for extended dramatic festivals to bring young and old together. Families brought food to the agora and camped out for days, both watching and participating in dramas that explored human behavior, natural history, and philosophy. The informal environment of the community theater allowed generations to learn together and communicate big ideas. Many historians believe that theater was an essential building block to the growth of democracy.

The model of a community theater is a great analogy for Family Science Nights like those described in this book. Lights and cameras may be optional, but the interactions that occur during an event like this are electrifying. The official program for Family Science Night is just the start. Intergenerational exploration allows learners of every age to discuss and ask questions. Like centuries of community theater, Family Science Nights emphasize communication. It's the talk that counts.

## Family Science Nights Are Rooted in Research

This is a book for pragmatic planners. But the events described in the pages that follow are firmly rooted in a significant body of research on the efficacy of informal education. Family Science Nights break down walls between institutional curricula and the world around us, creating a dynamic with enormous potential. Much of what we know about the importance of community involvement is summarized in *Learning Science in Informal Environments: People, Places, and Pursuits* (NRC 2009). This publication describes a number of ways that families interact outside of the official K–12 program for STEM discovery. Some examples are totally serendipitous: a chance encounter with an animal on the walk to school, a fishing trip, or a night sky observation. Sharing a great book or nature documentary might light a spark that ignites a passion for discovery.

But *Learning Science in Informal Environments* also documents the value of everyday and family science learning that occurs in settings "designed for learning ... museums, science centers, school experiences and the like" (NRC

2009, p. 127). The authors of the study admit that the distinction between everyday learning and learning in designed settings is "blurry and imperfect." A question that is raised in the classroom might be asked again on the walk home. A homework assignment might start a dinner-table argument or inspire a Saturday trip to a nature center. The possibilities are endless.

A Family Science Night combines the convenience of a resource-ready venue and the potential of a relaxing, welcoming social environment to generate communication. The literature that this publication summarizes includes a number of sound studies on the efficacy of outside-of-school intergenerational experiences. In schools, museums, and centers, gatherings that are scheduled outside of the normal school day encourage a special sort of communication. Through enhanced interest, time is multiplied because the conversations that begin at the classroom, museum, or café door never end. "Reflection on those experiences often extends after these experiences and is observed in future family activities" (NRC 2009, p. 96).

It's worth admitting at the outset that a Family Science Night will take time, talent, and even a little funding. But both research and experience show that family events are well worth the effort. Data document both higher in-school achievement for students who participate with their families and higher support for STEM in communities overall.

Like the community theater of ancient times, a Family Science Night can become the foundation of community spirit and democracy. In these settings, young and old share their cultures and ways of knowing in a way that respects diversity. By recognizing cultural perspectives outside of the narrow school curriculum, a family event can become an important tool to enhance equity and empower citizens in ways that go far beyond just learning science content. Learners in informal environments "experience excitement, interest and motivation … reflect on science as a way of knowing … think about themselves as science learners and develop identity as someone who knows about, uses and sometimes contributes to science" (NRC 2009, p. 4). STEM thinking empowers individuals and communities to become personally invested in problems and problem-solving in all areas of life. This book describes the sense of self-efficacy that results when learners of all ages are empowered.

## But Don't Be Fooled

When you see a great theatrical production, you may come away with the impression that it was easy to create. A great performer's chief talent might be to make it seem that he or she is "in the moment" and what you see comes naturally.

But we know that what occurs on the stage is the result of countless contributors and attention to endless detail. That's why this book will be invaluable.

Authors Donna Governor and Denise Webb have created a playbook that includes a tremendous number of resources and practical tips. This is the result of their combined experience and their pragmatic approach to making great events happen. From the "Overture"—an idea in the making—to the final applause, the path to success is well defined. Here you'll find dozens of details you might wonder about and a dozen more that you might never have considered. There are concerns you must address (such as funding, supplies, safety, and security) and things that might provide added value, from costumes to snacks.

Attention to detail not only contributes to success but also reduces the potential stress that a major production might otherwise generate. Knowing in advance that you've considered most of the eventualities will allow you to relax and enjoy, to become part of the community dynamics. And knowing that two experienced mentors like Governor and Webb are on your team will help you at each step of the way.

It's a pleasure to share with you this playbook for a successful Family Science Night. The set is prepared, the footlights are lit, the invitations are ready to be issued. Break down your classroom walls. There might be no "Tony Award" for the best Family Science Production, but there are many side benefits to these productions. What you will gain will multiply your educational space and your impact in a way that will amaze you. The unanticipated profits—from student achievement to community support for the school STEM program—will be great.

Follow the practical pathways to success in the pages that follow and then sit back and enjoy.

Lights, Camera … Science.

**Juliana Texley**
Past President
National Science Teachers Association

## Reference

National Research Council (NRC). 2009. *Learning science in informal environments: People, places, and pursuits.* Washington, DC: National Academies Press.

# Acknowledgments

We would like to acknowledge the wonderful support we've received from the students, parents, staff, administration, and our colleagues at Liberty Middle School, Coal Mountain Elementary School, and North Forsyth High School over the years that we've held our Family Science Night events. These events never would have been successful without the generosity of scores of people who have given of their time, talents, and resources to help make these programs successful.

We would also like to thank our husbands, Larry Morris (Donna) and Christopher Webb (Denise), who have encouraged and enabled us each year to pour so much into these events. They have often become willing participants in making the science happen.

# About the Authors

**Donna Governor** is an assistant professor of science education at the University of North Georgia. She has 32 years of K–12 experience in Florida and Georgia.

**Denise Webb** is an elementary science and engineering teacher at Coal Mountain Elementary School in Cumming, Georgia. She has been teaching grades K–6 since 1993.

# Introduction

Our experience with hosting Family Science Nights has been one of the most exciting aspects of our teaching careers. We launched a successful partnership organizing these events in 2013 while teaching in adjacent schools. We met serving on the board of the state science teachers association, and one of our discussions turned to our joint interest in planning a Family Science Night. Denise was trying to find a way to bring science nights to her elementary school, while Donna was looking for a way to get high school students involved in hosting science nights, after successfully organizing these events at the middle school level for the prior seven years. A productive and enduring partnership (and lasting friendship) came out of those early discussions. That partnership resulted in successful events for thousands of families at multiple schools over the past several years.

We both were products of the same preservice teacher program for elementary educators at the University of West Florida, although Donna completed the program 10 years prior to Denise. We lived and taught in different cities, and our paths wouldn't cross for more than two decades after Denise's graduation. When we began teaching, neither one of us dreamed that we'd eventually become teachers of science, as we both started our careers as elementary teachers. Yet today Denise is a STEM specials teacher at an elementary school where she provides hands-on labs and engineering activities for every student in the school on a weekly basis. Donna is now teaching preservice teachers at the university level, after having taught elementary school for 15 years, middle school science for 14 years, and high school science for 3 years.

For both of us, our early vision of a Family Science Night event was rooted in hands-on science events hosted by teachers at our respective elementary schools during the school day in the 1990s. How that experience evolved into organizing successful Family Science Night events, hosting thousands of students and their families each year, was very different for each of us. For Denise, these events started with teacher-led curriculum nights where she enthusiastically took charge of science activities. Denise saw students are naturally curious about the world, and science not only excited children about learning, but also inspired them to read and write more. Her interest gravitated to science based on the enthusiasm she saw when students engaged in STEM activities at these events.

Donna's path was more complex. After hosting a successful Star Party for her own middle school students in 2005 (with the help of a local astronomy club), the school parent-teacher-student organization asked her to host a similar event for the entire school in the following year. Although three attempts were made at holding the event, each was canceled due to cloudy skies and bad weather. The following year (2007), Donna tried again, this time with a different twist. Instead of relying on clear skies, she made the decision to include make-and-take activities, bring in a portable planetarium, and have the amateur astronomers prepared to "show and tell" their telescopes, should the skies not cooperate. Rather than relying on teachers to run the hands-on activities, she asked her eighth-grade students to take charge. Once students became involved, the plans began to snowball and more than 15 sessions ended up on the schedule. The skies cooperated, hundreds of students and their families showed up, and a new tradition at the middle school was born. These events continued to grow and evolve, until Donna made the move to teaching high school in 2013.

The partnership begun by Donna and Denise in 2013 started with a single school. Donna's former middle school students, now at the high school where she taught, asked to find a way to continue to sponsor Family Science Night events. Denise wanted to bring a Family Science Night to her school but needed volunteers to run the sessions. It was a match made in heaven. Our first collaborative event was held in January 2014 at Denise's elementary school, where several hundred students showed up with their families. In the same way that Donna's middle school nights evolved, the joint program took on a life of its own and continued to expand. Within two years, our high school students were hosting events at a half dozen local elementary schools for thousands of students and their families.

We want to make it clear, this book is not just for elementary teachers who want to host a Family Science Night at their own school. This book is for teachers of all levels—elementary teachers, secondary teachers, and teachers of preservice teachers. Through our collaboration, we found that Family Science Night events are great for our youngest students, but are important for older students as well. Middle school and high school students who are put in charge of running these events get as much, if not more, out of hosting Family Science Nights as the elementary students who attend. We've seen older students develop a sense of self-efficacy in science in ways that never would have happened in the classroom. If you are a secondary teacher interested in building a love of science in your students, this book is for you, as much as it is for the elementary teacher who wants to organize an event at his or her school.

Over the years, we've learned a great deal about how to host a Family Science Night. Throughout this book, we hope to share what we've learned and provide insights that can inspire you to organize an event in your school that truly excites learners of all ages about science.

# Section 1

## Producing the Event

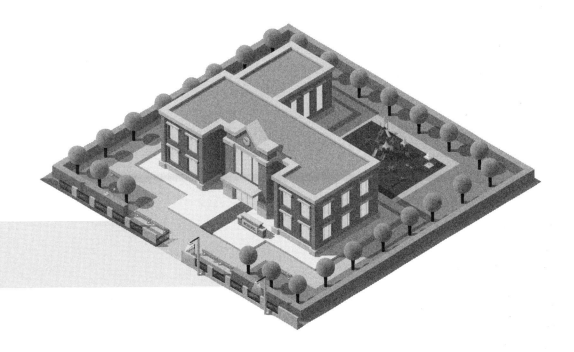

# Chapter 1

# Overture

## Overview

In this chapter, we will introduce you to Family Science Night events and discuss the benefits of holding an event at your school.

- The Case for Family Science Nights
- Building a Culture of Science
- Our Audience: Who Is This Book For?

- Evolution of Family Science Nights
- Summary

On March 22, 2007, aliens invaded Liberty Middle School in Cumming, Georgia. Approximately 40 students from Donna's eighth-grade science classes organized and hosted an Astronomy Night, which was to become our school's first Family Science Night event. Originally, the program was supposed to be small, with just a few telescopes borrowed from the local astronomy club scheduled to show students the stars. Hot chocolate and warm cookies were planned to keep things toasty and cozy. But, worried about the potential for bad weather, Donna scheduled a few extra activities to make sure the event could be held regardless of weather. She arranged for a portable planetarium from the local nature center and added a few make-and-take sessions to the program, such as making star clocks and planispheres.

Donna asked her eighth-grade students to help with the hands-on activities. That's when things exploded.

As soon as word got out, students started coming to Donna asking to participate. It was great to see students interested in hosting an after-hours event, and Donna started finding roles. Sessions were added to the plan, including activities on phases of the Moon, spectroscopy, exoplanets, and stellar evolution, as well as a "tour" of our Solar System. Students started coming up with additional ideas: How about T-shirts? What about a cookout? Where can we put an art show? What about music? By the night of the event, we had acquired over 1,000 Oreos, a dozen bags of marshmallows, inflatable planets, thousands of glow-in-the-dark stars, gallons of hot chocolate, and a dozen blow-up aliens. One very supportive parent even made over 1,000 candy stars for one of the activities!

We invited families from the local elementary school as well as those from our school. Students organized and hosted 17 different activities, including a performance by the school chorus, a local storyteller with myths about the night sky, and a hamburger dinner. Teachers were asked to participate as room monitors, and many chose to bring their own families. Six 25-minute sessions were held, with guests choosing the activities that interested them the most. The event went off almost flawlessly (although in all honesty, it was a bit of a whirl!). The students who led sessions were amazing! They knew their science, and they managed each event with a minimum of assistance from monitoring teachers. An estimated 600 people attended that first event. (See Figure 1.1.)

**Figure 1.1. Donna and Denise at a Family Science Night Event**

It was at the end of the evening that Donna realized what had occurred: a perfect coming together of informal science learning, student leadership, community support, and schoolwide excitement. Some of the comments made by the students who hosted the event will never be forgotten, such as "I couldn't believe how much fun it was!" and "I didn't realize how hard it was being a teacher!" It was immediately clear that this wasn't a one-time event, but the beginning of a new tradition that would evolve and grow. The event took on a life of its own and became a favorite for the school and community.

When Donna eventually transferred to teaching at a high school and met with Denise, a teacher at the neighboring ele-

mentary school, a new collaborative adventure began. We brought Donna's high school students to Denise's school to be ambassadors at our Family Science Nights.

## The Case for Family Science Nights

Family Science Nights such as these are considered an "informal" science learning environment—settings where children and adults can engage in and learn science beyond the classroom. The National Science Teachers Association's (NSTA's) official position statement on Parent involvement in Science Education states that "by sharing science experiences, parents demonstrate that learning is an important and enjoyable process. The parents also become more aware of the importance of science in their own lives" (NSTA 1994). We've found that science nights are one of the best ways to get students excited about and parents engaged with science. While participating in these events, students and their parents build science literacy, learn more about careers in STEM fields, and participate in scientific practices.

Family Science Nights are a relatively new phenomenon. They are first referenced in the professional literature beginning in the 1990s, which is when Denise and Donna were first exposed to them as teachers. The research tells us that Family Science Night programs can improve both knowledge and attitudes about science. For students who attend events, science literacy, general knowledge, and attitudes about science are improved (McDonald 1997). In a study published by Mike Watts (2001), the author found that benefits go far beyond impacting individuals who attend and extend into the community as well.

> As former PTO [Parent-Teacher Organization] president, I can confidently say that Science Night is one of the most important events of the year. It promotes family participation and the awareness of STEM. Everyone, no matter what the age, is highly engaged and leaves with a smile. Our PTO board always received positive feedback from the community on how much they enjoyed participating.
>
> —**Kim Hickman**
> *Former PTO President*

In the last decade, the National Science Foundation has published a number of reports on the importance of informal science learning, including *Learning Science in Informal Environments: People, Places, and Pursuits* (NRC 2009), *Surrounded by Science: Learning Science in Informal Environments* (NRC 2010), and *Identifying and Supporting Productive STEM Programs in Out-of-School Settings* (NRC 2015). Each of these documents discusses the importance of engaging students in learning outside the classroom environment for developing science literacy. In these reports, they also discuss the importance of learning in other informal learning environments, including museums, after-school programs, camps, and more.

We feel that student participation in a Family Science Night most closely parallels the experience of learning science in a "designed setting," as described in *Learning Science in Informal Environments* (NRC 2009). In these experiences, learners choose from multiple activities in which they can participate. The programs we present in this book fit this type of experience, as our programs are annual events that include dozens of different activities for students to choose from. Participating in these experiences can engage students, get them interested in science concepts, promote independent learning, and lead to deeper conceptual understanding of science content. For teachers, this means students who engage in learning science OUTSIDE the classroom are better prepared for learning IN the classroom.

What is written in the literature reinforces what we've seen with our own programs: participating in these events engages our students in unexpected ways. Young students ask better questions, discover new interests, and are better prepared for learning science in the classroom. Older students who run events suddenly see themselves as successful in science and as having the potential to be "good" at science. Families bond over experiences with meaningful discussions about phenomena and hands-on activities. For Donna's middle school students, the annual event became the high point of the year. With our high school collaboration model, the students who run the events not only discovered that they enjoyed working with children, but also found that their own sense of self-efficacy in science was improved. Students who never saw science as something they could "do" suddenly became active participants in the culture of science. Regardless of the age or role, these events seemed to bring out the science enthusiast in everyone who participated.

There are multiple models for running a Family Science Night. One model is to base activities on discrepant events that engage students and their parents through conceptual conflict (Lundeen 2005). Another model is for parents and students to work together at home to complete an investigation before participating in a culminating activity at a school-based evening event (Watts 2001). Some events are theme-based, such as the Astronomy Night project described at the beginning of this chapter (Governor and Richwine 2007). There is no one right way. Regardless of model, however, all events involve similar planning and preparation.

In this book, we will primarily discuss the models that we have experience with. The one described in Donna's first Astronomy Night is what we'll call a *session* model. This is much like a teacher conference, where concurrent sessions are scheduled and participants choose which activities to attend. The other type of model presented we call a *flow* model. In this type of event, multiple activities are set up, and participants move from one activity to another as they complete each activity. Both models include a variety of activities. However, in the session model, there is time to present a brief content overview to groups of participants. During a flow model, hosts must present content on a one-to-one basis, as attendees enter and leave each activity at different times. While the events look different, they require similar planning and preparation. Chapter 2 will discuss these models in more detail.

Regardless of which model you choose, planning should begin months before you want your event to occur. Selecting a date and time is dependent on a range of variables. Community events, sports, and even the onset of daylight savings time are variables that can make a difference in an event's success (more on this in Chapter 2). These events can be expensive (although not necessarily so), and raising funds to defray costs has always been part of the preparation for us. Part of planning an event includes deciding on formats, themes, and activities, recruiting volunteers, managing supplies, and arranging facilities. However, running a successful Family Science Night event is one of the most rewarding activities of any teacher's career, and this book is designed to help you make that process easier by learning from our experiences. Whether you are a classroom teacher, administrator, scout leader, or museum director, we hope you can benefit from our experience.

## Building a Culture of Science

Teaching and learning about science formally happens in the classroom. But learning about science involves interacting in the world around you. Since the beginning of the 21st century, science education has changed based on an ever-evolving body of research about teaching and learning. The release of *A Framework for K–12 Science Education* (the *Framework*; NRC 2012) and the *Next Generation Science Standards* (NGSS; NGSS Lead States 2013) has shaped what science education should look like in the classroom at the beginning of the 21st century. Engaging students in hands-on science experiences in a Family Science Night environment can address all three dimensions of learning identified in the *Framework*. "Three-dimensional" science activities engage students in both the content and practices of science, while emphasizing concepts that cross all scientific domains. Activities can be included in any Family Science Night event that span a variety of science and engineering practices, while engaging students in exploring patterns, relationships, system models, or causality. These three dimensions of learning are appropriate in and out of the classroom. They are specifically referenced in all the activities we present in the second half of this book and are important considerations when planning science activities.

National Academies of Sciences presents a parallel framework for learning in informal science environments in the report *Learning Science in Informal Environments: People, Places, and Pursuits* (NRC 2009). The report presents the six broad goals, or strands, to guide learning science outside the classroom. These strands are the following:

- Strand 1: Sparking Interest and Excitement
- Strand 2: Understanding Scientific Content and Knowledge
- Strand 3: Engaging in Scientific Reasoning
- Strand 4: Reflecting on Science
- Strand 5: Using Tools and the Language of Science
- Strand 6: Identifying With the Scientific Enterprise

For Family Science Night events, each of these important goals should be addressed to make sure that learners of all ages are engaged in meaningful science learning. These strands provide insight into how to engage learners beyond the classroom. A summary of these strands and how they apply to Family Science Night events follows.

## Strand 1: Sparking Interest and Excitement

Last night I took my two children to the science event put on by the high school students. I wanted to tell you that it was incredible! The science experiments were great, and every high school student that we encountered engaged with the kids and taught them the science behind the project. Please know how impressed I was with the organization of the Science Night and the overall science program. My oldest child is going into fifth grade, so we have a few years before he goes to high school, but he is definitely looking forward to being a Science Ambassador. Thanks for your great leadership to the school and community.

—**Chris Emmitt**
*Parent*

This strand deals with issues of motivation, excitement, and interest in learning science. It is probably the easiest of all goals to address when organizing a Family Science Night event for your school. There is extensive literature that discusses the importance of emotion in learning, concluding that students who are engaged and excited learn more and retain longer. But as every teacher knows from watching her own students, science activities spark excitement! Small children and their parents are excited whether they are exploring forces and motion with a water rocket, observing the Galilean moons of Jupiter, or dissecting owl pellets. Students who get excited about science in an after-school event are more likely to come to the classroom ready to read to find answers, write stories about their experiences, and engage in classroom science instruction. We've seen young children find a new passion for a topic because of an activity at a Family Science Night event, and older students who are hosting events suddenly feel as if they can be good at science for the first time in their educational career. Regardless of the age and role, engaging students in these events can spark interest and excitement. (See Figure 1.2.)

## Strand 2: Understanding Scientific Content and Knowledge

While we admit that science content cannot be learned in depth at a Family Science Night, there are many ways in which learners can improve their understanding of scientific content while attending an event. One way is by presenting scientific models that will help clear up misconceptions. For example, participating in experiences that model Moon phases will challenge misconceptions and help lay the ground work for classroom instruction. In this setting, younger children can experience a specific concept at an earlier age (such as Newton's Laws), and with the right questions, this will prepare them for future learning. Family Science Night events can also help by providing experiences that go beyond what classroom teachers can provide, such as the opportunity to observe planets at night. Providing experiences that activate new interests can result in later learning and deeper understanding. For example, when young students participate in a simulated fossil dig at such an event, they may not only ask important questions when they return home or to class, but might also be inspired to check out books at the school library to learn more about fossils. We've seen passions lit during these events that resulted in students seeking out more information. So while many students may not actually learn a great deal of content in short sessions, they can participate in experiences that will provide a base for constructing important knowledge at a later date, discovering a new area of science they might be interested in, or reengaging in content they may have forgotten. In *Learning Science in Informal Environments*, the report confirms what we found: Not a lot of

**Figure 1.2. Students Participate in a Robotics Demonstration at a Family Science Night Event**

content is learned in a single event. However, these events do improve students' potential for later learning (NRC 2009).

When older students are put in charge of a Family Science Night event, their understanding of science content and knowledge increases dramatically. Students who are responsible for presenting sessions at an event DO learn content at a much deeper level because they have to be able to teach the information to attendees. Our experiences include both middle and high school students who present sessions to younger students. In both our models—session and flow—middle and high students select a topic (with guidance), identify a hands-on activity, prepare a presentation, and then deliver content to attendees. These students had to learn the science and be able to explain it to younger students and their families. They couldn't always answer every question, but they certainly became more knowledgeable and had a strong motivation for developing a deeper understanding of the content.

> I have enjoyed being involved with the students and teachers taking science into the community of our local elementary schools as Science Ambassadors. The excitement of showing the students our experiments and watching them learn and see their reaction is priceless. The high school students and their leaders watch each child, wondering how far she or he will take their love of science. There might be an astronaut, chemist or doctor who starts at that one moment."
>
> —**Lorrie Angell**
> *High School Parent*

## Strand 3: Engaging in Scientific Reasoning

Scientific reasoning in this context correlates with the scientific practices identified in the *Framework* and includes observing, asking questions, predicting, experimenting, collecting data, and constructing explanations from evidence. These are at the heart of activities presented in a Family Science Night event. Discrepant events help students ask questions, engineering challenges require predicting and testing, and simple experiments involve data collection and interpretation. How much students engage in scientific reasoning will depend on the activities or stations you prepare. One of our favorite inquiry activities involves exploring potential and kinetic energy by making a "hall roller" (see p. 155 for this activity). This simple device can be constructed out of cardstock, straws, plastic cup lids, and rubber bands. The more you wind up the rubber band, the farther it rolls. Students can explore the relationship between potential and kinetic energy as they experiment with different amounts of elastic potential energy stored in the device. After constructing a hall roller, participants can collect data, manipulate variables, and measure outcomes as they compete to see which prototype is the best. By carefully selecting activities that encourage experimentation and asking

the right questions of attendees, Family Science Night events can help build an understanding of science and engineering practices.

## Strand 4: Reflecting on Science

Reflection, in this setting, deals with understanding science in a broad, social context as well as one's personal reflections on learning science. Understanding how science progresses, both historically and culturally, is an important component of scientific literacy in society. Reflecting on learning promotes meta-cognitive awareness, which is another way in which Family Science Night events can promote this strand. We see evidence of this strand as we hear from parents how much their children enjoyed the program every event. For our older students who host sessions at these events, they seek deeper levels of understanding to be able to answer questions from younger children. Integrating historical aspects of science can be accomplished either through specific activities or by selecting a theme based on building scientific literacy. For example, one of Donna's most memorable Family Science Nights was when the theme A Night of Discovery was used for the event. All sessions revolved around discoveries in science, focusing on a specific inventor, explorer, or researcher who made a historic contribution to advance science. Sessions included in this event covered the contributions of Annie Jump Cannon, Marie Curie, Gregor Mendel, Nikola Tesla, and more. Single sessions at nonthemed events that include references to history and culture can easily be integrated into any program.

## Strand 5: Using Tools and the Language of Science

This strand is perhaps one of the easiest to see in action at a Family Science Night event. Because of the nature of the program, students get to experience both tools and language that they might not be exposed to in the classroom. In our experience, many K–6 teachers don't have access to equipment that can be found in upper-level science labs. However, we've found that when organizing these events, it's easy to borrow the tools you need from a high school or ask for resources in the community. We've received donations of exam gloves from doctors, microscope slides from labs, and flowers from florists. Our local astronomy club has brought telescopes to our events, giving our students experience with the tools of science.

Perhaps the most important benefit from an event is exposing students to the language of science. Vocabulary used in various activities can connect with a wide range of science concepts. Asking, "What is your hypothesis?" before one activity, or introducing an "independent variable" during another, can help reinforce concepts students hear in class. More specific terminology can be introduced as students dissect flowers, manipulate Newton's laws in the Balloon-Powered Cars activity, and identify fossils from different geologic eras. Family Science Night activities introduce new words in meaningful and relevant ways.

## Strand 6: Identifying With the Scientific Enterprise

The focus of this strand is on "how learners view themselves with respect to science" (NRC 2009, p. 46). This is one of the strongest outcomes we have seen with holding Family Science Night events at our schools. Young children who attend these events become engaged with science and excited about new concepts. In our collaborative program, High School Science Ambassadors (more about that in Chapter 3), where Donna's high school students hosted events at Denise's elementary school, it was common for our elementary students to be overheard leaving saying, "I can't wait until I get to high school and can be a Science Ambassador!" These events make lasting impressions on the young children who attend and, we believe, foster a special relationship with science from their very first program.

It is with our older students—those who are responsible for hosting events—that we see the greatest shift in scientific identity. As these students take on the challenge of leading events for younger children, they develop a sense of confidence and love of science that they may have not developed in their classrooms.

As a parent of a college-bound student, I appreciate the leadership opportunities Science Ambassadors provide for students. Each group is given an activity. How they teach and share is their responsibility. High school students make decisions about how to handle difficult concepts, engage young children, and involve parents. They develop experiences and make memories that will last a lifetime.

—**Charlotte Stevens**
*Teacher and Parent*

These student hosts spend months preparing for events and take great pride in their work. In becoming experts in a single topic, they develop an identity as not only a learner but also teacher of science. When working with our High School Science Ambassadors, Donna loved recruiting those who didn't necessarily have a positive self-image of their ability to learn and "do" science. It often took some encouragement to get students to attend their first event, but once they did, they were hooked!

We've seen these students blossom and thrive when they become an expert that young children look up to. Additionally, as a high school teacher, Donna has seen a side of her students while participating in our Science Ambassadors program that was hard to find in the classroom. Students who don't seem to like science in class suddenly see themselves as being "good" in science, which, in

turn, improves their classroom interactions and motivation. For us, the transformation we've seen in high school students is one of the best parts of running the Science Ambassadors program.

## Our Audience: Who Is This Book For?

So, who is this book designed for? Really, anyone who wants to hold a family or community science event. It is written from our experiences hosting these events in public, K–12 schools (see Table 1.1). Elementary and middle school teachers can use the information included here to run events in their own schools. High school teachers might find this book a helpful guide to organizing a student program for future teachers or a science club to sponsor events at elementary schools, as we have done. Scout leaders and youth programs may want to use this book with clubs and organizations to help improve their events. Museum directors might find some useful tips for improving their outreach programs. Homeschool networks might also find ways to implement our ideas in community spaces. Just about anyone who wants to organize a science extravaganza outside the classroom could benefit from this book. In fact, we recently implemented ideas we learned while running Family Science Nights as part of a community science festival.

| Table 1.1. Who Should Use This Book and How to Use It ||
|---|---|
| **WHO** | **HOW** |
| **Elementary Teachers** | As a guide to organize and prepare a program for your school, using teachers or volunteers to host activities. If you wish to use students to host activities, we recommend you partner with a high school teacher to coordinate with and oversee students. |
| **Middle School Teachers** | As a guide to organize and prepare a program for your school, using students or volunteers to plan and host activities. |
| **High School Teachers** | As a guide to organize high school students to plan and host programs at local elementary schools. It is recommended you partner with an elementary teacher at the schools you intend to visit. |
| **Others** | As a guide to organize and prepare a program in an informal learning environment. This can include museums, community centers, children's activities at science festivals, and other venues. |

Depending on your current experience and program goals, you may only want to implement specific aspects of Family Science Nights or add certain features to an existing program. This book is intended to be a comprehensive resource, culminating from decades of combined experience.

## Evolution of Family Science Nights

One of the biggest considerations as you begin to implement your first Family Science Night event is that over time these programs tend to take on a life of their own and grow in unexpected ways. It is important to start simple and then expand as you gain experience. The vision we present here took us years to build. We recommend that you start with a small program and a limited number of activities, picking and choosing some of the ideas that we present in this book. In successive years, plan to expand your program to include more features. Sometimes our ideas have guided our programs, but more often than not, it's a student or someone else who has brought a great idea into play. A PTSA dinner held in collaboration with our event became a regular fundraiser. The art teacher added an art show with work related to the chosen theme. One year students included a coffee shop and listening room, performing original science songs (they actually produced a CD the first year the songs were performed). We adopted mascots, and the robotics club has presented demonstrations on multiple occasions. None of these ideas were ours, but they became important components of our programs over the years.

## Summary

Hopefully, by now we've built the case for why you should want to host a Family Science Night. Elementary students are engaged with science at an early age. Secondary students hosting Family Science Night events develop leadership skills and enhance their self-efficacy as learners of science. For families, these events provide a culture of science and build scientific literacy in the school and community. We've seen how they can provide a positive change for learners of all ages who participate in these events. In the next chapter, we'll talk about what you need to know to get started planning your own Family Science Night.

## References

Governor, D., and P. Richwine. 2007. Invite an alien to astronomy night. *Science Scope* 31 (3): 48–53.

Lundeen, C. 2005. So, you want to host a Family Science Night? *Science and Children* 42 (8): 30–35.

McDonald, R. 1997. Using participation in public school "Family Science Night" programs as a component in the preparation of preservice elementary teachers. *Science Teacher Education* 81 (5): 577–595.

National Research Council (NRC). 2009. *Learning science in informal environments: People, places, and pursuits.* Washington, DC: National Academies Press.

National Research Council (NRC). 2010. *Surrounded by science: Learning science in informal environments.* Washington, DC: National Academies Press.

National Research Council (NRC). 2012. *A framework for K–12 science education: Practices, crosscutting concepts, and core ideas.* Washington, DC: National Academies Press.

National Research Council (NRC). 2015. *Identifying and supporting productive STEM Programs in out-of-school settings.* Washington, DC: National Academies Press.

National Science Teachers Association. 1994. Parent involvement in science education. *www.nsta.org/about/positions/parents.aspx.*

NGSS Lead States. 2013. *Next Generation Science Standards: For states, by states.* Washington, DC: National Academies Press. *www.nextgenscience.org/next-generation-science-standards.*

Watts, M. 2001. The PLUS factor of family science. *International Journal of Science Education* 23 (1): 83–95.

# Chapter 2

# Writing the Script

## Overview

In this chapter, we'll discuss some of the initial steps you need to take to get started planning for an event and the decisions you'll need to make early in the process.

- Program Scale
- Approaching Your Administration
- Picking a Date
- Spaces
- Supply and Material Management
- Format
- Using Themes
- Timing
- Looking Ahead
- Safety Notes
- Summary

You know you want to hold a Family Science Night event, but where do you start? There are so many questions! When? Where? How? For us, getting started was the most challenging part of the task. Deciding to hold an event, getting our schools' administrations on board, and committing to the event were perhaps the biggest hurdles to overcome. Once the plan was in motion, it gained momentum and almost seemed to have a life of its own. As you begin to prepare for your first Family Science Night event, you should know that things will work out just fine. Not everything you want to include will happen, and things may not happen exactly as you plan, but great things will work their way into your event, including many ideas that you didn't expect when you started. Be flexible, and know that it will work out fine in the end!

Starting our first collaborative Family Science Night came out of a series of disconnected events that occurred in the fall of 2013. Donna had just moved to teaching high school and was approached by a group of her former students who wanted to continue holding Family Science Nights, even though they were now in high school. We were both on the board of our state science teacher organization, and during one of these meetings, we met and discovered we taught in neighboring schools: Denise was the STEM specials teacher at the elementary school adjacent to Donna's high school. Denise was looking for a way to facilitate a Family Science Night at her elementary school, and it wasn't long before we found our solution! Donna's students would come to Denise's school to host the Family Science Night event that she did not have the resources to implement on her own. It was a perfect solution for both Donna's students and Denise's plans. Denise took on the role of coordinating the activities, while Donna worked with high school students to participate as activity hosts. Administrators at both schools were consulted and gave their blessing, and planning began. By partnering Donna's high school with Denise's elementary school, we were able to work collaboratively on a project that benefited both schools.

We decided to call the program we created working with Donna's students High School Science Ambassadors and planned for a single event at Denise's school that January. We soon had about 40 students on board, recruited primarily from Donna's students, and scheduled an after-school club meeting. Our high school students formed groups, decided on activities, and began preparation for a midwinter event. Monthly meetings were scheduled after school for the fall semester, to plan, practice, and prepare for the event. Shirts were printed, name tags made, and materials ordered.

Denise joined the students for all the planning sessions to work with them to understand their activities and the science behind them. But she also had other responsibilities. She had to coordinate with her administration, identify classrooms to be used, and arrange for teachers to act as supervisors in each room. On the night of the event, there was a great deal of excitement as the High School Science Ambassadors arrived to set up. They used a map to find their assigned rooms, arranged materials and supplies, hung door banners for each session, and anxiously waited.

Families began arriving early, as they always do, and were moved into the cafeteria to wait while our ambassadors made final preparations. At the official start time, families were released to enjoy the event. For the next 90 minutes, parents, with their children in tow, moved from one room to the next and engaged in hands-on science. Donna and Denise circulated among rooms, but the students were well prepared, so there was little to be done. As at Donna's first middle school science night, the excitement was electric. Because a single night's program wasn't enough for the high school students, the Science Ambassador program soon expanded to other neighborhood elementary schools as well. Donna has since left the high school and is now working with preservice teachers at the university level, but the High School Science Ambassadors program has been sustainable by continuing with a new teacher assuming her role. This experience has been one of the best of our careers, and we have learned a great deal since then that can provide some insight into the logistics that will help you start planning your event.

## Program Scale

Before we go any further, it is important to clarify the importance of starting with a limited program for your first year. Donna's first year running events at the middle school level may have included more sessions than she initially intended, but it was not the grand-scale program it evolved into over the next seven years. The first collaborative program we implemented using Donna's high school students was a simple production, involving a dozen activities and a limited number of high school students. For both events, we knew immediately we had something special that would continue to grow!

What we've included here are ideas from more than a decade of experience. We are presenting a grand picture of all the things your program eventually CAN become. Hopefully, the ideas presented will help you build your program, adding new features with each iteration. We can't tell you which components to do year one, and which to add years two and three and so on. Each situation is different, with different needs and talents brought into the equation. You'll find some ideas work for you better than others. And you'll find new ideas we haven't considered that are ideal for your situation. So, start simple and keep it small. Take away a few ideas that you think will help you get started, and come back to look for ideas to help your program grow over time.

> I love Science Night! I volunteer every year to help. The excitement of the kids is contagious to everyone in the school. The kids love participating in the activities no matter what their age. It has become an event for the entire family. I believe this is the most looked-forward-to event every year.
>
> **—Joyce Hamby**
> *Teacher*

## Approaching Your Administration

As you begin planning your Family Science Night event, the first person to approach is your school administrator. If you are working with more than one school, then there are multiple administrations you need to approach. Usually there are no objections, but be ready to answer a few questions. When do you plan to hold your event? Where will you hold it? What resources do you need? It's generally a good idea to think through a preliminary plan with options, and consider some basic logistics first. It is even possible you may need to "sell" your event. As discussed in Chapter 1, there are many different reasons why your administration should support a Family Science Night event at your school. Any experience that brings families together to participate in science activities will promote learning. Interest in science and development of student identity as someone who can "do" science are also facilitated. When participating in informal science learning experiences with their families, students engage with greater interest and focus (NRC 2010). And if you are considering using our model of older students working with younger ones, pairing elementary schools with high school students builds community and boosts self-efficacy in science for older students who host events for younger ones.

## Picking a Date

You would think this would be the easiest of the logistics to determine, but picking a date is, in fact, quite a complicated dance. Scheduling a Family Science Night for the first semester isn't a good idea because it can take months of preparation. The holiday season is never a good time to hold an after-school event. After mid-March, spring sports are in full swing, reducing the potential for both volunteers and community participation. So, we've found that the sweet spot seems to be from mid-January to mid-March, preferably before the change to daylight saving time, when outdoor sports activities begin. During these months, there is less competition for participation in your after-hours event. Once you have an established program with a track record and a history of community involvement, you can be more flexible on the date. But for your first event, we recommend any after-school event be held in midwinter. However, you may find a different time of year works better for your community.

There are several other considerations related to timing that may affect your plan. If you plan to have an Astronomy Night, keep in mind that viewing celestial objects is best done on a first quarter Moon, when there is enough moonlight to observe craters and other lunar features, but not so much that the Moon's brightness washes out the ability to see stars and planets in more detail. Other considerations for scheduling include student programs and sports activities, as well as local festivals and events. We've found Tuesday nights to be successful with less competition for student time in our community, but you may find another day is best for your community.

## Spaces

If you are planning on holding your event at a school, activities will most likely be assigned to classrooms and other common spaces in your school. The spaces you decide to use will be dependent on the activities you decide to include. This isn't the time to assign specific spaces, but to think about the spaces that are available and potential uses for them. If you plan on having dissections, which are very popular activities, you'll want more of a traditional lab setup. If you want to bring in a portable planetarium, you'll need a large space with a high ceiling, such as the gym. Any type of rocket is better outside, but some rocket activities can be held inside, with high ceilings and plenty of floor coverings. Water-based activities, such as Foil Boats, will require a room with a sink and an easily mopped floor. Some activities will have space requirements that you won't necessarily think about until it is too late. Some spaces lend themselves naturally to great activities. The cafeteria can be used for a dinner event. The library makes a great location for staging a "coffee shop" and bake sale. Early in the planning process, make a site visit and look at each space, checking to see that furnishings and floor space are adequate and making sure there aren't any potential problems, such as class pets that could attract the attention of enthusiastic students at the event. We'll discuss spaces and how to use them more in Chapter 4.

## Supply and Material Management

Knowing what supplies and materials you'll need is dependent on the number and type of sessions offered. It's a good idea to think about organizing supplies and materials early in the planning process so you can have the necessary management system in place before you start acquiring needed items. You will want to create a master list that encompasses materials for all the activities and how they will be used. We found the use of a spreadsheet very helpful in managing supplies and orders. A spreadsheet can be sorted and re-sorted based on supply source or activity or alphabetized by materials. If you plan to take your event on the road for multiple events, as we did, you'll need a way to make sure supplies are adequate for the number of events and restocked as necessary after each one.

Although you'll need to purchase some materials, others you can obtain through donations. For example, owl pellet dissections and some of our other activities require surgical gloves, which often come from parents who work in the medical field. Local businesses, such as fast food restaurants, donate cup lids for Hall Rollers and straws for Rockets. A florist can provide flowers for dissection, and nearby grocery stores can supply cookies for various activities.

Parents also help by donating supplies. Be sure to keep notes in your spreadsheet on donations so that you can send thank you notes. You can also refer to these notes in future years when you find you need the same materials!

Think about how and where you will store materials prior to your event. In the case of a single annual event, supplies can be checked off and kept in their original boxes in a storage space until just before the event. Once we began working together, we needed a better system, because we held multiple events each year at different schools. For our Science Ambassadors program, we settled on plastic storage bins, with each activity getting its own box. On the outside of the box, we included a label indicating the activity it was for, and on the inside of the lid, we attached a list of the materials and the amount required for a single event. Students provided us a list of replacement items needed after each event so that we could restock and be ready for the next school. Some of the items you will need to provide in a "general" box of supplies include rolls of plastic table coverings to protect classroom furniture, markers, scissors, crayons, lots of tape for multiple purposes, cleaning supplies (mop, broom, buckets, and dustpans), and a first aid kit.

## Format

Probably one of the biggest considerations you'll want to make is the format of your event. As we mentioned in Chapter 1, although there are many different ways to organize a Family Science Night event, our experience is with two different formats, the session model and the flow model (see Table 2.1).

### The Session Model

In this model, events are scheduled to start at regular intervals, with attendees choosing activities based on interest, much like at a professional conference. Specific time blocks are determined for presentations and activities, with all attendees rotating to new sessions at the same time. In our events, six sessions were scheduled each evening, lasting for 25 minutes each with defined start and end times for each session. You'll note that even through sessions were 25 minutes long, the schedule shown (Figure 2.1) has them starting every half hour to allow for travel time. Because the target audience was upper elementary and middle school students, the presentations included a short content introduction (5–6 minutes) followed by a 15–20 minute activity. Group activities such as simulations, demonstrations, interactive games, and kinesthetic activities are easily facilitated with a session model, because attendees come and go from each activity at the same time. As an example, Figure 2.1 shows a table from Donna's 2010 middle school program showing the location of each session, using this model.

| Table 2.1. Event Models | | |
|---|---|---|
| CONSIDERATION | SESSION MODEL | FLOW MODEL |
| Movement | Everyone rotates at same time | Individuals move as necessary |
| Interaction | Large or small groups | Individual or small groups |
| Schedule | Defined start/end times | Varies by attendees' needs |
| Content Overview | Can present to large groups | Present to small groups/individuals |
| Activity Length | Determined by schedule | Can include activities with varying time requirements |
| Group Activities | Easily facilitated | Challenge to include |

(The blocks of time shaded for each session represent "planning periods" provided for each group of hosts to take a break.) One of the advantages of this model is that content material can easily be presented to groups of participants before engaging in an activity. However, because there are specified session times, attendees can only participate in a specific and limited number of activities and have to make choices about what sessions to attend.

## The Flow Model

In this model, there is no specific schedule and individuals can take as long as they need to complete one activity before moving to the next one. One advantage is that this model allows for more individual interaction, as attendees arrive at each activity at different times. Hosts explain the science behind the activity individually to the young participants as they work through each activity. One of the drawbacks to using a flow model is that it is difficult to include group activities such as simulations, demos, and interactive games. We've found that the flow model works well for younger

### Figure 2.1. Session Model Overview

**Session Overview**

*Check the list to see what time each session is offered and the location. If there is a block "grayed" out, the presenters for that session are on break.*

**Sixth Grade Hall**

| Session Title | 6:00 | 6:30 | 7:00 | 7:30 | 8:00 | 8:30 |
|---|---|---|---|---|---|---|
| Old School Communication | 433 | 433 | 433 | | 433 | 433 |
| Popeye's Steamboat Adventure | | 434 | 434 | 434 | 434 | |
| Hydroelectricity! | 437 | 437 | | 437 | 437 | 437 |
| Energy Stories | 438 | 438 | 438 | | 438 | 438 |
| Bill Nye Energy Movie | 442 | 442 | 442 | 442 | | 442 |
| Fossil Fuels & Footprints | 443 | | 443 | 443 | 443 | 443 |
| Energy Transformations in Toys | | 447 | 447 | 447 | 447 | 447 |
| The Power of Wind | 448 | 448 | 448 | 448 | 448 | |
| Static Electricity | 455 | 455 | 455 | 455 | | 455 |
| Sound & Waves | 459 | | 459 | 459 | 459 | 459 |
| Flying High with Hot Air Balloons | 465 | 465 | | 465 | 465 | 465 |

audiences and allows for students to engage at an age-appropriate level. We anticipate our activities to take 5–15 minutes each, to match the attention span of younger students, and our programs include an estimate of the time it should take to complete each activity (see Table 2.1, p. 23).

For both formats, we recommend planning more activities for your event than attendees can possibly participate in. For us, this is done in part to accommodate all the older students who want to participate in hosting these events. However, the major benefit of having more activities is that each is less likely to be overcrowded. Our events are attended by hundreds of students and their families, and the only way to accommodate everyone is to have more sessions or activities than can be attended by a single participant. While we do not have attendees sign up for activities in advance, a variation of our model could include schedules for different groups of students to follow. A strict rotation order can be established if activities are limited, or a lottery system might be implemented for determining participation. We found that allowing choice works well, and when activities or sessions are full, we close classroom doors and instruct participants to choose another activity but come back later. Activities in high demand, such as the planetarium and dissections, are ticketed events, with a small fee and a reserved time (Chapter 4 has more information on this).

## Using Themes

Another decision about your program's format will be related to whether you want your event to have a theme. A theme is a great way to introduce novelty each year. Themes we've used in the past include Astronomy, Oceans, Biomes, and Energy. Donna's favorite theme was A Night of Discovery. All sessions were related to specific scientists and their discoveries. Themes can be rotated or chosen to fit specific situations. For example, you might choose a theme related to a particular science discovery in the news or to a local initiative.

Themes can provide a unique feel for each of your events, and there are a number of creative ways to integrate the theme into activities and the atmosphere of the entire event. Programs, T-shirts, and advertisements can all make use of the theme. Family dinners held in conjunction with the event can take on a unique name, such as the "Safari Café," the "Captain's Galley," or the "Space Port Diner." Music, art, and stories can encompass the theme as well. For the Science Ambassadors program (high school students hosting elementary schools), we made a decision not to theme our events, but this may change as the program continues to grow. See Figure 2.2 for an example of themed artwork.

For your first event, you'll also want to think about the focus and format of your activities. If you will be using a session model, what does a typical session look like? Do you want students to have something to take with them from a hands-on session, such as a rainstick? Or possibly a trinket used in an activity, such as a shell that was used for classification? Will you want all activities to use a standard lab sheet? If you are using a flow model, how will activity hosts manage students who are entering and exiting the activities at different times? What format do you want for your program, if you use one at all? Tips about making these decisions are found throughout this book.

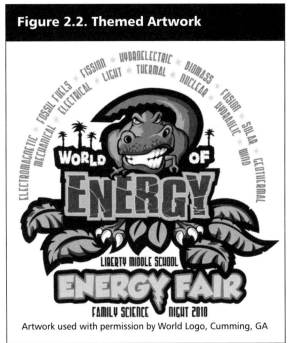

**Figure 2.2. Themed Artwork**

Artwork used with permission by World Logo, Cumming, GA

## Timing

One of the things we haven't discussed yet is how long your event should last. For an elementary event, students will arrive early and leave early. Yet starting your Family Science Night before parents get off work limits the number of students who can participate and is an important equity issue. We've found that for our elementary schools, holding events for 90 minutes between 5:00 and 7:30 allows for parents who work to bring their children, yet still wrap up for earlier bedtimes. It also permits our High School Science Ambassadors to leave with ample time to complete their own homework. For Donna's middle school events, the timing ran later, with a PTSA dinner starting at 5:00 and sessions starting at 6:00 and ending by 9:00. Often families didn't stay for the entire event, and attendance fell off after the 8:00 session. But, because students are older, there were always plenty of students and their families participating until the very end, including many who were later arrivals. Alternatively, you may choose to hold your event on a Saturday, which is not something we've tried yet at our schools. However, at our recent community Science Festival, we used a flow model to host a family science activity session on

Science Night is one of the most looked-forward-to nights out of all our events. Parents have said, "This is the one night we never miss. Our kids would not let us!" It brings families together, no matter what level of knowledge they have in science. They feel comfortable and excited to learn about all the STEM activities that are offered.

—**Kim Fox**
*Assistant Principal*

## Checklist 2.1.

### Planning Timeline

Based on our experiences, here's our suggested timeline for planning. If you plan to hold your event in the fall, you will need to accelerate the timeline.

**September**
- ☐ Get administrative approval
- ☐ Set a date
- ☐ Decide event format
- ☐ Decide who will run the activities
- ☐ Recruit activity hosts (students, volunteers, or teachers)

**October**
- ☐ Plan events and activities
- ☐ Determine spaces to be used
- ☐ Recruit adult monitors
- ☐ Compile supply list
- ☐ Invite related groups to participate
- ☐ Arrange ticketed activities and sessions
- ☐ Hold fundraisers
- ☐ Find advertisers for program
- ☐ Arrange bus transportation for hosts if using older students for events in elementary schools

**November**
- ☐ Order supplies
- ☐ Schedule practice sessions for activity hosts
- ☐ Review any student-developed materials
- ☐ Assign additional roles (e.g., photographers)
- ☐ Prepare a draft program

**December**
- ☐ Organize and prepare materials
- ☐ Rehearse presentations
- ☐ Finalize program

**January**
- ☐ Print programs
- ☐ Hold final activity practice

a Saturday morning, and we found plenty of parents who were willing to bring children out for a morning of science activities.

Regardless of your start and end time, be sure that you give activity hosts ample cleanup time. Check with your administration about availability of custodians in advance, and keep an ongoing conversation with them about what is required for cleanup. We encourage our activity hosts to take pictures of the room before they set up and again after they clean up to make sure the room is returned to the original arrangement. Hosts are expected to take out trash, wipe down desks, and for the really messy activities, sweep and mop the floor. This factors into our schedule, as we add a half an hour at the end of the event for our hosts to clean up before they are released. This attention to neatness has kept us on good terms with the teachers who allow us to use their rooms, and with our schools' administration.

## Looking Ahead

It is important to note that whatever model or format you choose for your Family Science Night event, it won't look exactly like anyone else's. Each situation is slightly different and each community has unique needs. Your event will also evolve and grow in unexpected ways as people with good ideas come on board. You'll learn after your first year what works and what doesn't, and you'll adjust accordingly. You may find some of our ideas work well, and some may not. There will be community resources that will bring a unique feel to your program that you will want to tap into. If you make your Family Science Night an annual event, over time your program will establish a reputation that will increase attendance and bring new resources and opportunities to you.

To help you plan your event, we've provided a timeline (Checklist 2.1) that you can use to guide you as you move from inception to production. This timeline is based on holding your event early in the second semester and should be adjusted if you plan to hold your event at another time. You'll note that some of the items on our list haven't been discussed yet, such as programs and fundraising. You may or may not need to raise funds, and you may or may not decide to have printed programs; however, we have found these elements have enhanced our events. We'll talk about them in more detail later in the book. We can't emphasize enough the need to start small and build, adding more components each year, so not everything on our list will apply to your program, but it can give you some insights about how to build your events in future iterations.

## Safety Notes

Throughout this book are hands-on activities and demonstrations that help foster the learning and understanding of science during Family Science Nights. The reader will find safety notes and safety statements that help make the event a safer learning experience for students, teachers, and other participants. In many cases, personal protection equipment (PPE) such as safety glasses or goggles, nonlatex gloves, and aprons are required. Sanitized safety glasses and indirectly vented safety goggles must meet the ANSI/ISEA Z87.1 D3 safety standard. When dealing with hazardous chemicals, consult with the safety data sheets prior to doing the activity. Make sure there are appropriate engineering controls; portable eyewash, ventilation, and so on. Always do a "dry run" of the activity prior to doing it in front of students. The safety procedures and use of PPE must be followed based on legal safety standards and better professional safety practices. Teachers should also review and follow local policies and protocols used within their school district and school, a chemical hygiene plan, Board of Education safety policies, and so on.

General safety protocols include but are not limited to the following:

1. Remind all students that personal protective equipment (sanitized indirectly vented chemical splash goggles meeting the ANSI/ISEA Z87.1 standard, nonlatex gloves, and nonlatex aprons) are to be worn during the setup, hands-on, and takedown segments of activities that note such equipment is required.

2. Remind students eating or drinking is prohibited.

3. Always review and model appropriate safety procedures prior to working with sharps to reduce or eliminate the risk of cutting or puncturing skin.

4. Some students may have allergies or have immune system deficiencies. For this reason, nonlatex gloves must be worn.

5. Wash hands with soap and water after completing the lab activities.

For additional safety information, check out the National Science Teachers Association's Safety Portal at *www. nsta. org/safety*. Check out resources such as the elementary, middle and high school level safety acknowledgment forms for other safety protocols.

Be aware that conditions of actual use of activities and demonstrations may vary, and the safety procedures and practices described in this book are intended to serve only as a guide. Additional precautionary measures may be required. NSTA and the authors/reviewers do not warrant or represent that the procedures and practices described meet any safety code or standard of federal, state, or local regulations. NSTA and the authors/reviewers disclaim any liability for personal injury or damage to property arising out of or relating to the use of this book, including any of the recommendations, instructions, or materials contained therein. Selection of alternative materials or procedures for these activities may jeopardize the level of safety and therefore is at the user's own risk.

## Summary

This chapter has been an overview of the basic ideas you'll need to address to start planning your event. The following chapters provide more details for you with specifics about different components that can enhance your program. In the next chapter, we will address special roles—including hosts, photographers, and materials managers—that volunteers can assume to make your event more successful. Special touches, such as materials bags, programs, and passports, will be discussed in later chapters. We'll talk about how to make your event more festive and how to get the community involved. Finally, we'll talk about what to do on the big night and how to keep from going crazy. We hope you will find useful information in the chapters that follow.

## Reference

National Research Council (NRC). 2010. *Surrounded by science: Learning science in informal environments*. Washington, DC: National Academies Press.

# Chapter 3

# Casting Your Event

## Overview

In this chapter, we will discuss the various tasks and the human resources that you'll need for a successful event.

- The Case for Working With Student Volunteers
- Specific Roles

- Recruiting the Cast
- What's Next?

Regardless of when and where you decide to hold your Family Science event, you will need a small army of volunteers and helpers to make it happen. There are dozens of tasks to be handled in running an event, and deciding who does what is important. As events get larger, you'll find you need a larger volunteer force. How much you do yourself and how much you assign to others is up to you. Of course, the first year, when your event is smaller, you will probably want to handle most of the logistics yourself, but you'll find it best to shift to an oversight role within a few short years as your program grows.

In this chapter, we will cover human resources: what roles you need to fill, and where to find the help you need. Because one of our primary strategies is to put students in charge, we will emphasize the role middle and high school students can take in your event and how to manage them. If you do not plan to use students but instead have preservice teachers, faculty, or parents as activity hosts for your event, the same information will apply about roles, responsibilities, and oversight. You'll have to modify the context, but we expect the information will be relevant and useful regardless. What follows is a summary of the tasks that need to be handled and how to cast these roles.

## The Case for Working With Student Volunteers

For us, in both the middle school and high school setting, using students to host sessions was critical to our success. We could not have found the human resources required for the events we wanted to organize any other way. In the case of Donna's middle school events, using students happened quite accidentally, while for the joint events Donna and Denise held together, student involvement was by design. Regardless, the importance of student volunteers cannot be overstated, as it not only gave our events the workforce necessary, but also provided both middle and high school students with a leadership experience that was unparalleled in any other school activity.

You might recall from our first chapter that Donna's initial event at the middle school involved about 40 of her students. Of the 17 activities included in the program, 15 of those were run by students in the advanced science classes. Using the session model, groups of three to five students took responsibility for and managed each activity as hosts. For each session, they were instructed to present 5–10 minutes of introductory content, followed by a 15–20 minute hands-on activity. In addition to the student-led activities, there were invited sessions, such as the local astronomy club, the portable planetarium, and a storyteller. For invited sessions, students were asked to chaperone, rather than host the activities on their own. Regardless of role, for our students, "Family Science Night became the most anticipated activity for any student in our advanced/gifted classes. Students were proud to be a part of it and as soon as one Family Science Night ended, the underclassmen immediately began planning their session for the next year" (Sherri O'Hara, Teacher, Liberty Middle School).

> I enjoyed watching students pursue their interests so that they were truly engaged in learning. I also really enjoyed the autonomy and responsibility that students had while preparing their presentations. Also, the fact that the presentations were open to the public led to high levels of accountability for students. I base a lot of my pedagogy on the lessons learned from Family Science Night.
>
> —**Chris Sartain**
> *Teacher*

In the middle school setting, participation in our annual Family Science Night event was tied to classroom expectations in advanced content classes where planning time for student presentations was facilitated in class. Student activity hosts had the responsibility of not only presenting but also planning and organizing their sessions. Months ahead of the event we introduced the theme and a general outline of the event. Class time was used on a biweekly basis for several months ahead of the event. Students rehearsed and revised their presentations in advance, and the advanced language arts teacher even used student presentations as an opportunity for authentic assessment in her curriculum. The annual Family Science Night event was truly a student-driven and student-led experience.

Later, when we began to collaborate on our joint Family Science Night events, the lessons Donna learned managing middle school events helped as we transitioned to using high school students to run programs in the elementary setting. We already knew the value of putting students in charge and allowing them to take the lead in hosting an event. Because participation was extracurricular and not tied to a class activity, recruitment, and preparation was a bit more challenging. Instead of allotted class time to plan, our activity hosts had to be willing to use time outside of school to prepare. Any time you interact with older students in an environment outside of class, you find things you would otherwise overlook, and our experience was no exception. At these events, the class clown became an impressive role model for elementary-age students, high school students who "didn't like" science completely changed their position, and students who had trouble completing homework became responsible and prepared presenters. Each year, word of our Science Ambassadors Club spread and additional students volunteered to participate, facilitating bigger programs. In both settings, we found engaging students as activity hosts to be rewarding and beneficial to everyone involved. Students gained leadership skills, while we were able to provide better events for families that attended.

> As an English language arts teacher, Family Science Night provided a wonderful opportunity for students to apply and showcase many of the skills learned in language arts—creating fluid written presentations, technology/media skills, demonstrating mastery of writing conventions, creating step-by-step directions, preparing written handouts, and oral presentation skills.
>
> —**Sherri O'Hara**
> *Teacher*

If you are going to use older students as hosts, we have learned a few things that will help you manage them more effectively. First, regardless of age, start with a signed commitment from students that includes parental approval (see Appendix B [p. 174] and Appendix D [p. 178]). You need to make time requirements clear on the consent form. Make sure students know to clear their calendars for the date of the event and to make arrangements to be picked up at the close of the events. Students volunteering as hosts will need to be picked up approximately half an hour after your program ends to allow for time to clean up. If you are holding the event in your own school, you won't need to transport your hosts to the event, but if holding the event elsewhere (for example, a local elementary school), you may need to arrange bus transportation for your hosts from their home school to the school where events will be held. Check with your district for any special transportation requirements that may apply.

Be clear about expectations for attendance, but also plan for about 10% of your students to be absent from the event. Illness, transportation issues, and unplanned schedule conflicts will result in a number of well-meaning students being unable to attend the event. By assigning student hosts to work in groups of three to five, a single absence will not affect the group's activity at the event. Make sure you set behavior expectations for your student hosts, and be clear about consequences for inappropriate behavior. For us, students who misbehaved during any planning activity or demonstrated they could not be trusted during an event were dismissed from the program. See Checklist 3.1 for a list of volunteer requirements.

Give your students a timeline, and check with them at every opportunity to make sure they are on target to have their session or activity ready on time. One of the most important things we've learned is that students will take just as much time as you give them to prepare. Whether it's six weeks or six months, they will find a way to be ready on time, and not one minute earlier. We both have had times when we worried about an apparent lack of progress, only to find our student hosts ready and fully prepared on the night of the event. Expect a lot of last-minute hustle and bustle, regardless of how much time hosts have had to prepare. They will be ready, and if not, only you and the students will know things aren't exactly as hoped. The families attending won't know what isn't perfect.

Not everything works out perfectly when students are in charge, and hiccups happen, but sometimes problems can provide an important life lesson for student hosts. Because the purpose of using students to host Family Science Nights is to have a student-led and student-driven

## Checklist 3.1.

### Student Volunteer Requirements

When working with students, make sure that they complete the following activities:

- ☐ Attend an orientation meeting
- ☐ Obtain parental permission to participate
- ☐ Agree to participation requirements
- ☐ Commit to event date
- ☐ Prepare activity or presentation
- ☐ Attend work sessions to prepare
- ☐ Submit a supply list of required supplies
- ☐ Practice and rehearse the presentation or activity
- ☐ Make arrangements for transportation home
- ☐ Decorate and set up space as necessary for activity or presentation
- ☐ Clean up activity space after event

event, you may even need to let student hosts make mistakes. While we don't suggest allowing pandemonium to erupt, we sat back and allowed students to problem solve even if their activities were not well thought out. We've found that activities can be revised for future use, and that our students learn and grow from experiences that are less than ideal. With enough activities, your event will still be a success even if a session or two isn't, and mistakes can provide an opportunity for student hosts to mature and develop valuable leadership skills.

> Sometimes things didn't go as planned, and we had to tweak the instructions to make them work better.
>
> —**Rachel Jennings**
> *Former Student*

By now, we hope we've presented a compelling argument for using students to host your Family Science Night event. Advanced middle school students as well as high school students can effectively plan, prepare, and manage activities with guidance and support. We've found that these experiences facilitate leadership skills, provide an opportunity for authentic, integrated learning, and build self-efficacy in science for the student hosts. So, what do you do if you are an elementary teacher? Approach your administration first, and then reach out to the principal of a nearby middle or high school for potential colleagues to work with. For us, and our students, it has been a great partnership and has helped build a sense of community within our school district.

## Specific Roles

In this section of the chapter, we provide an overview of the specific jobs that need to be filled and the responsibilities associated with each. Not all these roles will need to be included in your first event, but over time your event will grow and require more support.

### Activity Hosts

Whether you use students, teachers, or parent volunteers, the role of activity host is the most important one to fill to hold a successful Family Science Night. You will need a small army of volunteers to manage the activities you want to include in your event. For each activity you schedule, you'll want two or more people to host. When working with adult volunteers as activity hosts, you might find two people

> Science Night is such a wonderful event and it grows each year. In addition to being amazed at the excitement in the elementary students' faces, watching the high school ambassadors' faces as they interact with students creates an atmosphere of learning beyond compare. The high school ambassadors serve as true leaders in their excitement working with the students. Their passion is evident and supports passion in the elementary students. This is an incredible opportunity to foster the vertical teaming as a cluster of schools.
>
> —**Kimberly Davis**
> *Principal*

per activity is sufficient. When working with students, the number of students you assign to each activity will vary. For your most dependable workers, a group of three may be sufficient. For other groups, you may want four to five students to manage each activity.

Once you decide the basic logistics for your event, determining activities and assigning activity hosts will be your next major challenge. You can take a proactive approach by determining which activities you want included and assigning them based on what you know about your volunteers. Alternatively, you can let hosts determine their own activities, either by choosing from an activity bank or by coming up with their own ideas. Early in the planning process, stress proper safety, and continuously review all activities, regardless of the source, to make sure they follow accepted legal safety standards (e.g., fire codes, OSHA [Occupational Safety and Health Administration] standards) and better professional safety practices (e.g., National Science Teachers Association safety protocols).

How much responsibility your activity hosts take on depends on you. You can prepare all the materials yourself and have everything ready for them to lead activities on the night of the event, you can turn over virtually all responsibility to your hosts, or you can find a balance in the middle. If you are allowing hosts, whether adults or students, to prepare activities and materials, you will need to make sure you review the content and oversee preparations. Check materials for accuracy and look for misconceptions that should be avoided. Make sure your hosts understand the science behind the activity and are able to explain it to the children and families that will be attending their activities (see Figure 3.1 for a photo of student hosts).

**Figure 3.1. Student Hosts From the 2008 Oceans Extravaganza at Liberty Middle School**

Rehearsing the activities is a must for hosts. They should practice conducting the activity with others before the event. They should know the content related to the activity and be able to answer questions. Hosts should expect that something will go wrong, no matter how much they practice, so prepare them to be flexible and creative problem solvers. Having extra materials on hand is recommended, and if hosting events on multiple nights, expect variations in facilities and attendance. See Checklist 3.2 for a list of activity host responsibilities.

Prior to the event, be sure to discuss the importance of respecting the space your hosts will be using. We ask that our hosts decorate the room to make it a fun and engaging space, and of course, sometimes furniture has to be rearranged to accommodate specific activities. To make sure that the room is returned to its original arrangement, we ask hosts to take pictures of the room with their cell phones before starting and check to see that everything is put back in its place when they finish. We use table covers for some activities, and at the end of the event, we ask our hosts to wipe down desks, take out any trash, and sweep or mop the floor if necessary.

Coaching student hosts includes not just preparing them in advance of an event but also reviewing the event in a reflection activity afterward. You'll want to discuss what went well, what didn't, and what they would like to improve. We've seen both middle and high school student hosts make great improvements in their presentation skills from one event to the next. One comment we always hear is that students who host these events walk away with a greater respect for what their own classroom teachers do on a daily basis. These reflective conversations are an important part of program sustainability.

## Checklist 3.2.
### Activity Host Responsibilities
There is some overlap on this list with the checklist for student volunteer responsibilities; however, this list is more specific to the activity host role.

- ☐ Determine activity
- ☐ Review content for mastery
- ☐ Write abstract for program
- ☐ Prepare presentation, if applicable
- ☐ Practice presentation or activity
- ☐ Request necessary materials
- ☐ Organize materials and bring to event
- ☐ Determine and arrange decorations for your session
- ☐ Set up for the event
- ☐ Create a fun atmosphere for your session
- ☐ Interact with children at event
- ☐ Clean up
- ☐ Request additional materials, if applicable, for future events
- ☐ Store or return materials

## Monitors

There are several critical places that will need adult supervision, and the number and placement of monitors will be influenced by whom you use for activity hosts. If using student hosts, you will want one adult monitor present in each room where activities are taking place. This is critical for security and safety reasons, to protect the hosts from any potential accusations, to oversee behavior, and to make sure presenters are acting professionally. While we don't encourage our attendees to show up without adult supervision at any of our events, nonetheless,

## Checklist 3.3.

### Monitor Responsibilities

If assigned to an activity room:

- ☐ Introduce yourself to the activity hosts
- ☐ Monitor attendance to avoid activity exceeding capacity
- ☐ Handle any disruptions from parents and children attending the activity
- ☐ Report any problems with activity hosts to the event coordinator
- ☐ Dismiss activity hosts at the end of the event after the space has been cleaned up

If assigned to a common area (e.g., hall, cafeteria):

- ☐ Monitor attendees moving from activity to activity
- ☐ Handle any disruptions from parents and children attending the activity
- ☐ Report any problems to the event coordinator

some do. We recommend using certified teachers, para-professionals, or other school personnel who are certified or have passed a background check to supervise when students or nonschool personnel are hosting activities. See Checklist 3.3 for a list of monitor responsibilities.

For activity monitors, one of the most important jobs is to make sure the participants do not exceed the capacity of the room for the hosts that are managing the activity. When too many attendees show up for a session or activity at the same time, monitors should either stand at the door and ask attendees to come back later or close the door and post a sign indicating, "Activity Full, Come Back Later." Allowing too many participants into a session will create stress for the hosts and anxiety for impatient parents. It will also stretch resources beyond capacity and create a potentially unsafe situation. You should plan to have monitors stationed in the hallways and other common areas. Hall monitors can answer questions, assist parents and children in finding activity locations, guide them toward less crowded activities, and suggest return times for the most popular sessions. For legal reasons, parents and other adult volunteers should be asked to find a designated event supervisor or certified school personnel to handle any behavior situations.

## Greeters

Regardless of how large or small your event, you should station one or more people at the entrance to your event as greeters. The tasks required for volunteers in this position include distributing programs, answering questions, and providing directions. You may even wish to distribute bags printed with the current theme logo as we did (see Figure 3.2), or use disposable bags so that your attendees can have something to put their make-and-take science activities in. Another activity you might want to include at your reception area is a raffle. Donations made by advertisers, including gift certificates and merchandise, make great raffle prizes for fundraising. If collecting surveys, be sure to include a box for attendees to return them as they exit the event.

When using a session model for your event, the greeters should direct guests arriving between sessions to a common area until the next session starts. One great way to make this waiting area a pleasant experience is to set up an art show with student work provided by the school's art teacher, a demonstration by the school's robotics club, a coffee shop where parents can browse the school's book fair, or another activity that will be appreciated by your community. More on how to integrate these types of activities will be covered in Chapter 6.

One of the activities we include at our reception area is handling tickets for any special activities. While we want our events to be free of charge to engage students in science, some activities are so expensive that the only way to include them is by selling tickets for participation in those specific activities. It should be noted that ticketed activities should be a small fraction of the activities included in any event. Preselling tickets can be logistically challenging, but greeters can easily sell tickets at the reception area. Arrange for a cash box with ample change to start the event, and provide pre-printed tickets that specifically note the activity and time. More information on what types of activities you might want to charge for and how to handle ticketing is provided in Chapter 4. Even though greeters don't have as much responsibility as hosts do, they are the first people guests come in contact with, and it is important that the students or adults assigned to this role are helpful and maintain a positive disposition. See Checklist 3.4 for a list of greeter responsibilities.

**Figure 3.2. Family Science Night Bags**

## Checklist 3.4.

### Greeters

- ☐ Decorate the entrance to set the mood for the event
- ☐ Become familiar with the event so that you can answer questions
- ☐ Distribute programs and bags to attendees—usually one per family
- ☐ Facilitate pickup or selling of tickets for special activities
- ☐ Greet attendees with a smile
- ☐ Clean up the entrance at the end of the event

## Checklist 3.5.

### Chaperones

- ☐ Decorate the space the guest will use to set the mood for the event
- ☐ Welcome the guest upon arrival
- ☐ Offer the guest water or refreshments
- ☐ Help the guest set up
- ☐ Monitor and help with traffic in and out of the guest's space
- ☐ Clean up the entrance at the end of the event

## Chaperones

For many of our events, we like to invite and include guests. In the past, this has included a storyteller, the local astronomy club, and guest speakers. For each of these activities, we've assigned a chaperone to assist our guests with logistics and crowd control. Our chaperones have the responsibility of creating an inviting atmosphere in the space the guest will use, welcoming them when they arrive, seeing that they have what they need, and helping monitor the flow in and out of their space. They are also responsible for cleanup and making sure the space used by the guest is returned to its original condition. Students work especially well in this capacity as it allows those who might not want to take a leadership role the option of making an important contribution to the program with less responsibility. These positions are also great ways to introduce new volunteers to your program. See Checklist 3.5 for a list of chaperone responsibilities.

## Photographers and Videographers

This is one important job you will want to make sure is covered! Family Science Nights are exciting events that bring together schools and communities. Photos and videos from these events can be used to document all the tasks volunteers are expected to do and provide a compelling narrative for why they should be involved. Pictures from our events have been highlighted in the school yearbooks and shared in the school's lobby as a digital slideshow. Administrators may want to use these materials to document STEM activities as part of their school's culture.

There are a few requirements for our photographers. Be sure to tell your photographers that you want hundreds (really!) of pictures taken and that you want photos from before, during, and after the event. You want plenty to choose from. Photographers should take pictures of both activity hosts and attendees, showing active partic-

If you want to involve teachers and ask them to volunteer for your Science Night, engage them just as you would your students. Prepare your sign-up list by session or activity and allow teachers to sign up together for something that interests them. This does two things: (1) it gives teachers an idea of what to expect for their session, and (2) it allows them the opportunity to work together with a peer. If you are short volunteers, be prepared to personally invite teachers to sign up

—**Colby Counter**
*Teacher*

ipation. They should also plan to spend some time after the event reviewing and editing the images for composition. They should eliminate poor-quality photos and select only the best of any series of images taken at the same time.

A video of the event is also important for a number of reasons. We've used these videos as part of the school's newscast, both to report on recent events and to recruit for future years. Video can portray the excitement of an event in ways that still photography can't. You'll want to make sure that your videographer knows that you expect the final product to be an edited film of a specific length, usually two to five minutes, with opening and closing credits. These are skills many middle and high school students now have, and software to produce these videos is common and accessible.

One of the most important considerations with photography relates to using pictures of students and children from your event. Generally, it is expected that pictures will be taken at school functions and used in yearbooks and for school resources. Videos are often made and shared in school newscasts. So, if you are using pictures and video for use only within your school and program, you may not need to obtain specific permission to photograph students at your event. Check with your school to find out your district's policy and any requirements. Regardless, post a sign in the reception area and include information in your pro-gram to indicate that photos will be taken and that if parents don't want their children to have their picture taken, they should let greeters know upon arrival. Provide wristbands for those attendees who should not be photographed, and make sure your photographers know to check for a band before taking a guest's picture. If you will be using images beyond the school (e.g., for publicity, advertising, or publishing), you will need to get special permission to photograph your attendees. If this is the case, we recommend stationing a specific volunteer at the reception area for the purpose of obtaining releases as guests arrive and distributing wrist-bands as necessary. See Checklist 3.6 for a list of photographer and videog-rapher responsibilities.

## Checklist 3.6.

### Photographers and Videographers

☐ Take photos/videos of all preparation activities

☐ Obtain parent permission to photograph students if necessary

☐ Confirm that your equipment is fully charged prior to the event

☐ Make sure you have backup batteries on hand

☐ Be sure to get images/video of all activities

☐ Focus on participant engagement in your photos/video

☐ If taking photos, review and edit photos for composition

☐ If producing a video, edit and add credits

☐ Submit finished product/portfolio to coordinator within one week

## Managers

While this is not likely a role you will want to hand over to others your first year, you'll find that as the event grows, you'll want help managing the details. Because our Family Science Nights have always been student-led events, we've handed several managerial tasks over to our high school student managers (see Figure 3.3). There are a multitude of tasks that can be included for this position, determined by each unique situation. Some of the activities our managers have taken over for us include supply management, T-shirt orders, program development, and activity coordination. As events get bigger, overseeing materials and supplies becomes a major task. Not only do they need to be ordered but also

**Figure 3.3. Science Ambassador Student Manager With Denise**

they should be inventoried and organized. Our managers have been a great help with this task. We'll talk more about materials management in Chapter 5. Another responsibility that managers may be able to take over preparing programs (see Chapter 4).

Our managers also take up a mid-level position between us, as the organizers, and the activity hosts. They have the responsibility of verifying and collecting supply lists, distributing materials, updating us on progress, and handling special needs for the hosts. On the night of the event, managers are tasked with helping with setup and circulating through all the activities during an event to make sure that things are running smoothly. Finally, managers verify that all spaces are returned to their original condition at the end of the event. Additional tasks can be assigned, depending on your needs. Be sure that the job is assigned to volunteers, either students or adults, who are responsible enough to handle the demands of the position. See Checklist 3.7 for a list of manager responsibilities.

## Supporting Roles

There are a few additional roles you may want to cast for your Family Science Night event. First, if you live in an area with a bilingual population, you may choose to have interpreters on hand. Interpreters can translate your program, answer questions, and provide general information to attendees at your event.

Another potential role to fill is that of announcer. Be sure someone knows how to use the PA system in your building and can make both planned and unanticipated announcements. There are always general announcements to be made at any event, and you will need to encourage attendees to head out at the close of the event. If using the session model, you'll need to make announcements about when one activity period ends and another begins. When casting the role of the announcer, make sure the person has a good "teacher voice" that will be heard. Family Science Night events are often busy and loud, and announcements need to be heard over the noise.

## Checklist 3.7.
### Managers
- ☐ Identify activity groups that require assistance for the event sponsor
- ☐ Assist in ordering supplies
- ☐ Inventory and organize supplies as they arrive
- ☐ Track volunteer hours of all participants
- ☐ Assist in preparing event programs
- ☐ Oversee and distribute common materials (e.g., table covering, tape, brooms)
- ☐ Circulate at the event and assist in supply issues
- ☐ Advise event coordinator of any issues that need to be addressed
- ☐ Verify that all spaces used during the event have been cleaned and returned to the original (or better) condition
- ☐ Perform other tasks as assigned

## Recruiting the Cast

By now, you've seen that there are many different roles that can be filled. Other than activity hosts (and monitors when using students as hosts), all are optional. The first year of both our programs, there was no additional helper beyond our students. We handled the logistics and our students were activity hosts. As the program grew, we found we needed more support and required more roles to be cast. Additionally, parents, teachers and other community members became increasingly involved.

So how do you get the volunteers you need? Start by recruiting students you know. Over the course of several years, we found our students volunteered for a number of reasons, such as resume building (jobs as well as colleges), volunteering, working with a favorite teacher, earning extra credit, or collaborating with friends on a project. Offer community service hours to students, and make sure you have a system to accurately log and track their hours because many colleges now require service hours from students for admission. Many of the students who have participated in our science ambassadors program don't have another niche for extracurricular involvement at school, and our club provides a unique opportunity for this population.

Recruiting teachers can be more challenging. With our model of using students as hosts, there was little for teachers to do except oversee the students. We found that each year it was easier to recruit teachers to help. As the program grew, so did the enthusiasm and excitement. Teachers offered to take over specific tasks to be a part of the program. They often brought their families to these events. Some teachers asked to be involved because of the enthusiasm they saw participating students have. It was participation by one of these teachers that allowed the program at the high school level to continue past Donna's departure from the district.

Don't forget, your parents are a tremendous source of help for running a successful program. Approach your parent-teacher organizations for volunteers as well as financial support. Use relationships you've established with parents to start building your volunteer network. Our students' parents have offered to help in more ways than we could ever have expected. Parents make donations of materials and supplies, provide cash for incidentals, make homemade cookies and treats for activities or fundraisers, and generously volunteer their time both before and after events. Their businesses have become our program's sponsors and advertisers. It was the generosity of parents that allowed our program to flourish and made it financially sustainable.

## What's Next?

Now that you've decided on what help you need and who will handle what roles, you'll want to start planning for the little things that will make your program successful. While you won't want to implement all the roles we've discussed during your first year running an event, you'll find some of the ideas we mentioned helpful as your program builds over subsequent years. Next, we'll look at how to best use the facilities you will need for your event.

# Chapter 4

# Building the Set

## Overview

In this chapter, we present information about planning for the spaces you'll use and the activities you'll assign to each space. We'll also discuss creating a program for your event.

- Common Spaces
- Assigning Spaces

- From Planning to Program
- From Set Design to Props

Running a Family Science Night is one of the most rewarding experiences of our careers. We watched our programs grow at each level, eventually becoming one of the most highly anticipated activities of the school year. Chances are your first event will only include the essentials (activities run by your hosts), but over time, you'll want to expand the ways to engage families at your event. As the event becomes an established part of your school's culture, you'll find more people want to be involved, and you'll find new ways to broaden the impact and enrich the program. Our students returned year after year to help, even after they had moved on from the school. We feel that was due in part to the atmosphere that we created at each event. So it was a real positive feedback loop: The more cool stuff we integrated, the more help we needed, the more help we received, and the more cool stuff we could include! In this chapter, we will discuss building your "set." We'll include some ideas for using common spaces to build a richer event, assigning locations for activities, and creating a program for your event.

**Figure 4.1. Reception Area Signage**

## Common Spaces

Family STEM events seem to be living, breathing, and growing things. It has been our experience that they start with the basics and evolve. They seem to take on a life of their own, and both programs we've managed grew in unique ways. While each has its own culture and climate, there are some common characteristics in how common spaces are used. For example, you want the entrance to your event to be a lively, exciting place that sets the mood from the minute attendees walk in the door. However, you may or may not want to include food options at your event. Regardless of how your event evolves, you'll want to consider how to use common areas to build your program. See Figure 4.1 for an example of a student-made sign for a reception area.

### Reception Area Activities

Making sure you set the right atmosphere for your guests as soon as they enter the event is important. Use decorations to make it seem festive and welcoming! This is where you will want to distribute your programs, but you may also have other activities set up here that let your attendees know they are in for a night of science fun! When possible, ask your greeters to make the reception area an exciting place to help set the mood of your event. If you have a theme, use that to decorate accordingly. You might also make the entrance area look like a science lab right out of a television set or the bridge of a spaceship. Use songs with science themes to set the mood and have greeters wear lab coats.

One activity we keep in the reception area is our Science Surplus Store. We found that often we had leftover materials from one year that weren't needed the next. These included rubber balls, magnets, and other small items. These are sold during later years for a small amount—

> The one thing that stands out to me is that former students would return each year to participate. Former students would show up, and it would be not only a science event but also a small reunion, with many asking if they could help. Families looked forward to bringing the siblings of current students, year after year.
>
> —**Robert Dodd**
> *Teacher*

usually a quarter. If you want, you can even create bags of materials for activities and sell those, providing an easy way for parents to do activities at home with students, which might bring between fifty cents and a dollar.

As discussed in Chapter 3, if you have ticketed activities, this is the place for attendees to purchase or pick up their tickets. While we want all of our Family Science Night events to be free, we found some activities so expensive or popular that they created problems as families got frustrated with long lines for the planetarium or owl pellet dissection sessions. By selling tickets for these high-demand, expensive activities, we found we could add more experiences at our events and better manage the crowds. Tickets for the portable planetarium and dissections (e.g., owl pellets and flowers) are sold in advance and at the door, and for specific times. We recommend keeping the cost small; we charged two dollars. But, because these sessions were ticketed events, attendees were guaranteed a place in the session with no wait, allowing for them to better plan their experience. While the cost was minimal, it was enough to cover the expenses involved for including potentially expensive activities. When tickets for a particular activity were gone, the activity was sold out and attendees would need to make other choices. Depending on your demographics, you may or may not wish to include ticketed events.

Selling tickets in advance requires some work but is easy enough to manage with a good spreadsheet. Several weeks prior to the event, send a flyer home to parents with information about the event. Include a list of activities with information about ticketed events. Additional items that can be presold include T-shirts and meals, but more on this later. Figure 4.2 shows the preorder form used at Donna's middle school events. While we often host hundreds of families, only a handful purchase tickets in advance; most wait and obtain theirs upon arrival. When the order form arrives with payment, the parents' and child's names are recorded in a spreadsheet with specific information about the event. A cutoff date is given for preordered items. Preprinted tickets are assigned to each family for their purchases and put in an envelope to be picked up at the reception area. Be sure to assign a responsible adult to be in charge of the money, and have a cash box and plenty of small bills for change ready in advance.

## Figure 4.2. Preorder Form

EVENT PRE-ORDERING FORM

☐ $5.50 for Prepaid Dinners - Total Number: _____
Indicate number of each: ___ Hamburgers ___ Cheeseburger ___ Hot Dogs

☐ $15 for T-shirts - Check one for each shirt ordered - Total Number: _____
☐Youth Small ☐Youth Med ☐Youth LG ☐Youth XL
☐Adult Small ☐Adult Med ☐Adult LG ☐Adult XL ☐Adult 2X

☐ $2 for each planetarium ticket (advance cost): Total number: ____ X $2 = _____
Indicate Time Preferences for Planetarium for first, second & third choice times:
___ 6:00 ___ 6:30 ___ 7:00 ___ 7:30 ___ 8:00 ___ 8:30

TOTAL AMOUNT ENCLOSED: _____

Make checks payable to the school & turn in to Donna Governor

ALL PRE-PAID ITEMS WILL HAVE TICKETS/SHIRTS AT THE DOOR FOR PICK UP THE NIGHT OF THE EVENT

EMAIL DONNA GOVERNOR FOR MORE INFORMATION OR IF EARLIER CONFIRMATION DESIRED FOR PREORDERS *AFTER* PAYMENT SENT IN

*ALL PREORDERS FOR T-SHIRTS MUST BE RECEIVED BY MARCH 1*
*FULL PRICE MEALS AND PLANETARIUM TICKETS AVAILABLE AT THE DOOR*
*PREPAID ORDERS WILL BE TICKETED AND AVAILABLE FOR PICK UP AT THE DOOR*
*T-SHIRTS AVAILABLE FOR PICK UP AT THE SCHOOL ON 3/18*

### Dining and Concessions

Having a family meal attached to your event can really be beneficial. While we don't recommend you add this task to your plate, your school may have a group that would like to sponsor a meal. At the middle school level events, Donna's Parent-Teacher-Student Organization (PTSO) hosted a benefit dinner each year. Some of the elementary schools we've taken our High School Science Ambassadors to have also added a dinner in coordination with the event. These dinners can help enrich the event by providing community interaction. Usually a local burger or sandwich shop can provide meals for a per-person cost, and the price to attendees should allow the school to make a dollar or two profit on each meal sold. For Donna's middle school, this was a major source of income to fund the event. Because the PTSO took responsibility for managing the event, there was no oversight required. If you want your families to have more time for the activities, open your "diner" a half hour before your activities start. If you are hosting the event at another school, you might suggest that school's administration host a dinner to raise funds to help cover the cost of activities for your event.

Another food-related activity that we found successful is a coffee shop. This is an ideal place for families to browse the school's book fair, if scheduled in conjunction with the Family Science Night. Be sure to include hot cocoa for children as well as coffee for adults. A bake sale can be added to this space as another source of income to fund the event. Don't forget that you can even add an activity to this space. For example, one year some of Donna's students performed science songs they had written for the event as entertainment. Another year, a parent offered to do face painting in this space, with donations accepted to help fund the event. The coffee shop and bake sale can be organized by parent volunteers or students. You might find you want to keep your coffee shop open a half hour after your activities end to give families something to do on the way out and provide time for hosts to clean up.

A final refreshment-related activity you might wish to provide is a "hospitality suite" for your activity hosts to take a break. Donations of snacks and beverages can be placed in this location, and a break can be scheduled for each group. With the session model, break times should be staggered so that different activities are closed at different times. In the flow model, even just a 10- to 15-minute break will give hosts a chance to use the facilities and catch their breath. Including a hospitality suite sends a message to your hosts that they are valued and appreciated.

## Assigning Spaces

When designing your "set," keep in mind specific needs for each activity. Probably the biggest mistake we ever made in assigning rooms was the first year that we had making ice cream as an activity. This recipe requires mixing milk, sugar, and vanilla in a plastic sandwich bag and placing the small bag inside a larger bag with rock salt and ice. Wisely enough, we stationed this activity near water access, but we didn't consider the distance to the cafeteria, where the ice machine was located. Our poor student hosts spent much of the evening hauling ice back and forth from the cafeteria to the activity room! From this, we learned to look more closely at the needs for each activity, as you will quickly learn to do. In Section 2 of this book, we've included some information about special facilities required for activities on our Teacher Tips pages.

Activities that need water should of course be placed in a room with a sink or, at the very least, near a bathroom. Foil Boats, Cartesian Divers and Catch the Wave all require water nearby. Messy projects and food activities, such as Bubble Olympics and Ice Cream, will require water for cleanup. These activities should not be stationed in a room with carpeting. If a refrigerator is necessary for storing materials that need to be kept chilled, place the activity close to the teacher's lounge or bring a cooler.

If you plan to include dissections of tissues or organisms, such as squid or cow eyes, you will need a standard laboratory room, complete with lab tables, venting, and proper equipment. Occasionally, we allowed for squid dissection, but only with strict protocols for safety and supervision in place. Dissections of simpler items such as flowers and owl pellets only require limited lab equipment such as goggles, nonlatex gloves, nonlatex aprons, and a handwashing station with soap and water. One of our favorite activities during an astronomy-themed event was Alien Dissection. This activity involved preparing alien-shaped molds with gelatin and fruit. Attendees learned about the possibility of life on other worlds before cutting into their alien specimens. Another year, attendees performed "brain surgery" by removing a brain tumor (grape) from a gelatin brain mold. These activities should be strategically placed near water and close to a refrigerator to store the "specimens."

Some activities use space in unique ways. Our portable planetarium, a staple at the middle school events, requires a high ceiling, as do rockets made from film canisters. An online simulation activity can be explored if you have a computer lab available. Paper airplane experiences call for more floor space, as do roller coaster activities. At our astronomy events, the local astronomy club sets up just outside a door near the darkest part of the schoolyard. Should the sky refuse to cooperate, telescopes can be brought inside so that these guests can explain how they work, in place of the viewing activity.

When assigning spaces, also pay careful attention to traffic flow. You will want activities spread out and not crowded together on the same hall. Too many attendees in the same hall can create chaos and frustration. However, you don't want your activities so spread out that you are running a marathon just to check in on your hosts. We recommend assigning spaces based on need rather than allowing hosts to choose their spaces. Sometimes student hosts will request specific spaces and may ask for a change of venue to have better exposure, to have specific furniture, or to work with a favorite teacher. In general, we don't recommend entertaining changes without a good reason. It leads to hurt feelings and a steady stream of requested revisions. We also suggest that you talk to the teachers and personnel who are willing to allow their classrooms and spaces to be used to make sure that they understand your needs and that you understand their expectations. In the past, we've had misunderstandings about students putting up posters and other decorations, and complaints about tables that were returned to a position a few inches off from the original location. We try not to use rooms that belong to teachers who are less than enthusiastic about the program and place the messier activities with teachers who have the most tolerance for chaos. Teachers should be encouraged to stow or cover items they don't want touched before the event. You might provide classroom teachers with a list of activities prior to assigning spaces and allow them to choose the activity that is assigned to their classroom. Sometimes teachers will choose to hold certain activities in their rooms due to a positive rapport or relationship with certain students. Teachers who are parents of student hosts are a good example of when this is beneficial. Finally, if possible, make sure your hosts check out their activity space as soon as possible and let you know if they see any potential problems.

## Figure 4.3. Program Matrix

Choose from a variety of exciting sessions. Pick the ones that look like fun, but make some alternate choices. Once a session is full, the door will close and you'll need to attend another. But not to worry – there is a lot of great fun to be had no matter where you go!

| Activity Schedule | 5:30 | 6:00 | 6:30 | 7:00 | 7:30 | 8:00 |
|---|---|---|---|---|---|---|
| Deep Sea Dinner & Art Show Opens at 5:00 | Cafeteria | Cafeteria | Cafeteria | Cafeteria | | |
| Celestial Navigation in the Portable Planetarium | Media Center | Media Center | Media Center | Media Center | Media Center | Media Center |
| **Connections Hall (500 Rooms)** | | | | | | |
| Timing the Tides Gizmo Lab | 505 | | 505 | 505 | 505 | 505 |
| Oceans Online | 509 | 509 | 509 | 509 | | 509 |
| Music of the Deep: Band (GYM) & Chorus (MP room) present special music sessions | Gym | Gym | | | Multi-Purpose Room | Multi-Purpose Room |
| Fish Biology 101 | 526 | 526 | 526 | | 526 | 526 |
| **Sixth Grade Hall (400 Rooms)** | | | | | | |
| Waves Across the Ocean | 437 | | 437 | 437 | 437 | 437 |
| Ocean Currents | 443 | 443 | 443 | 443 | | 443 |
| Food Webs of the Kelp Forest | | 459 | 459 | 459 | 459 | 459 |
| Nautical Knots | 465 | 465 | 465 | 465 | 465 | |
| **Seventh Grade Hall (300 Rooms)** | | | | | | |
| Mapping the Ocean Floor | 333 | 333 | 333 | | 333 | 333 |
| Underwater Archaeology | 334 | | 334 | 334 | 334 | 334 |
| Movie & Popcorn: Living Sea | 342 | 342 | 342 | 342 | 342 | |
| Where in the Ocean Do I Live? | 343 | 343 | | 343 | 343 | 343 |
| Bioluminescent Critters | 344 | 344 | 344 | 344 | | 344 |
| Scuba Diving | | 359 | 359 | 359 | 359 | 359 |
| Squid Dissection | 365 | 365 | 365 | 365 | 365 | |
| **Eighth Grade Hall (200 Rooms)** | | | | | | |
| Beachcombing for Sea Shells | 207 | 207 | 207 | | 207 | 207 |
| Storytelling: Ocean Myths | 212 | 212 | 212 | 212 | 212 | 212 |
| Pirates & Their Treasure Maps | | 213 | 213 | 213 | 213 | 213 |
| Adaptations of the Coral Reef | 229 | 229 | | 229 | 229 | 229 |

## From Planning to Program

Of course, programs are essential to every event, from year one on. You'll definitely need to provide information about what activities attendees can participate in and where to find them. Our first event programs were nothing more than a single sheet of paper with a list of rooms and activities. A matrix (Figure 4.3) provided a quick overview of activities, times, and where to find them. The reverse side of our single-sheet program provided additional information about the activities, including a one-

or two-line summary of the activity and hosts' names. For example, this was the description for the session Adaptations of the Coral Reef: "Animals have adapted to all kinds of environments, and in the coral reefs you'll find all kinds of ways animals have learned to survive. A coral reef friend could even follow you home!" Of course, we had to use small fonts to get all the information on one sheet of paper, but it was sufficient and provided the necessary information for attendees.

As our event grew, so did our programs! Our programs became multipage documents that included a great deal more information. The middle school program added a number of extra features, while our programs for the elementary events added information about each activity (Figure 4.4). The activity pages in Section 2 of this book are based on the format we used in these programs. We wanted not only the name of the students and a description of the activity, but also information about the science behind the activity, required materials, and a how-to so that it could be completed at home. It was our hope that by providing this information, families would engage in the activities they couldn't do at the event when they returned home or repeat their favorite ones they visited.

Additional pages also went into our programs. One important feature to include in your programs is a map (see Figure 4.5, p. 50). Showing the location of each activity can help attendees plan their experience and find the sessions they want to attend.

## Figure 4.4. Sample Activity Page From Program

### Activity: Musical Harmonicas

**Student Ambassadors:**

| Science Behind the Activity: | Sound is created when something vibrates. On our kazoo blowing air makes the rubber band vibrate against the stick which then moves air molecules to create sound. |
|---|---|

**What you need:**

- [ ] 2 popsicle sticks
- [ ] 3 rubber bands
- [ ] 1 straw
- [ ] Scissors
- [ ] Markers

**How to:**

1. Cut 2 straw pieces, about 2 in. long.
2. Place one rubber band around one popsicle stick - the long way.
3. Put the two straw pieces at each end under the rubber band (see image).

**Instructions & General Information**

4. Place the second stick over the first stick and straw assembly.
5. Wrap the ends with the remaining rubber bands.
6. Decorate your kazoo with markers
7. BLOW!

**Estimated Activity Time:** 8 minutes

If you are holding your event at just one location, then you can embed the map in your program. If you are holding your events at multiple locations, eliminate the map in your printed program and provide a map insert for each specific event. Rather than creating a map of the building from scratch, scan the fire evacuation map your school uses and start adding information with a graphics program. The map will help make sure that you are using spaces wisely during planning. A map will also help you manage traffic flow on the night of the event and will

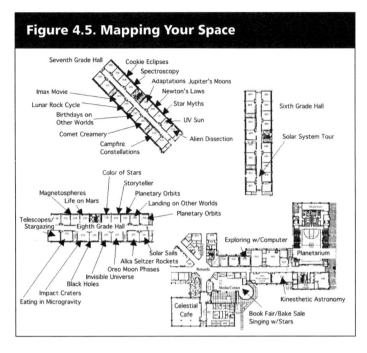

**Figure 4.5. Mapping Your Space**

help attendees find their sessions quickly. The map that is shown in Figure 4.5 is from the event held in 2011 at Donna's middle school. This was the fifth year of events there, and the program had grown from roughly 15 sessions at our first Family Science Night to more than 30 in 2011. You'll notice that the astronomy theme had rolled back around after several years of other themes.

Other pages you may wish to add include a table of contents, a letter of support from your principal or superintendent, a survey, and a passport. The survey allows you to get some feedback about what sessions went well, what didn't work, and what suggestions attendees have for improvement. In the past, we've also asked for specific information about student hosts who were well prepared and professional. The passport is a useful addition that can be used to document the activities students participate in. By providing student hosts with stamps, attendees can collect evidence of the activities they attend. Teachers can use this evidence for extra credit in the classroom or as part of a reward system. Probably the most important reason to use an expanded program is that it allows you to sell advertising space as a way to raise funds to support the program. We'll have more information about advertising in Chapter 6.

Creating an extended program is easy using a word processing document. Use a landscape format on regular copy paper so that each sheet of paper will provide four pages of text. Use large margins that will produce a page that fits on half a sheet of paper in landscape format. You can save your program as a PDF document for electronic delivery. When you print your program pages, make sure they are arranged in the correct order so that your program prints front and back, then make as many copies as you need. You can obtain a long-arm stapler to assemble your program. We recommend using the logo or artwork from your event on the cover. In the first few pages of the program, include general information about your program, a letter of support from your administration, and a table of contents. Figure 4.6 shows how to create your program layout.

## From Set Design to Props

In this chapter, we've talked about spaces you'll use. The spaces you'll need will be based on the activities you decide to include in your event. We've included ideas for special activities that make use of additional spaces, such as the cafeteria for community dinners and the media

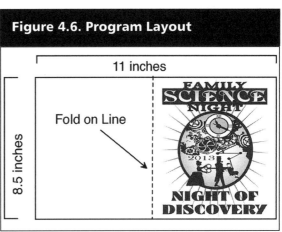

Figure 4.6. Program Layout

11 inches

8.5 inches

Fold on Line

center for a possible coffee shop. Suggestions have been made for activities to include in your reception area as well. We've also provided some tips for making an event program. Next, we'll explore how costumes and props can make your event more exciting for your attendees!

# Chapter 5

# Costumes and Props

## Overview

In this chapter, we will discuss a variety of extra touches you can include to give your event extra pizazz, such as the following:

- Branding Your Event
- Costumes
- Props
- Sprucing Things Up
- Managing Your Costumes and Props

To make your event extra special, you want to dress up the party! The use of costumes and props has the potential to invigorate your Family Science Night with a festive atmosphere of fun and excitement. For us, this is where the magic occurs. Shirts, hats, signs, decorations, and mascots have all benefited our programs by setting the mood for a fun-filled event. These little touches create a sense of belonging and community for our student hosts and a festive world of science for our attendees. Our student hosts eagerly await the arrival of each new T-shirt design and often plan elaborate decorations for their sessions. From the minute they walk in the door to the end of the evening, we try to make each aspect of the event special for our guests. In this chapter, we'll share what those extra touches are and how to facilitate a similar atmosphere at your own event.

## Branding Your Event

The use of specific artwork for each event started Donna's first year running a Family Science Night and continues to this day. Most T-shirt companies, such as the one Denise and Donna have been using for the past decade, can create a custom artwork design for use on shirts, signs, programs, and advertisements. If you plan to order shirts or other products, there usually isn't a charge for the artwork; the cost of the design will be wrapped into the cost of the goods. Otherwise, you may be asked to pay a small fee to cover the design work. The art that is produced will become important to your event and be used in multiple ways. Of course, it will be used on shirts, but you can also use the art on promotional materials, programs, bags, and signs. Each year, we provide our vendor with a general concept to use to draft a logo. We try to limit our designs to one or two colors to keep the cost of printing shirts down. You might also want to use a graphics program to change your design to grayscale before finalizing it, to see how well the colors contrast in black and white, as programs will most likely not be printed in color. Just to be on the safe side, make sure your vendor doesn't mind your using the artwork to promote your event. You may wish to provide advertising space in your program as a perk for the vendor's cooperation.

**Figure 5.1. Student Host Leah Troop in Her Ocean Adventure T-shirt**

## Costumes

For our student hosts, the T-shirt design has always been one of the most eagerly anticipated aspects of our programs (see Figure 5.1). Whether using themes for an event or not, the artwork is presented to students in an exciting introductory session. For hosts' shirts, we have their role printed on the back, although the term often changes with the theme. Astronomy Night hosts are "staff," but Biome Safari hosts are "guides." Our ocean-themed events use the term "crew," although the term "host" can also be used. When purchasing shirts in large quantities, the cost is usually under $10 each. We use donations and income from fundraisers to help students who are not able to afford the cost of purchasing their own shirts. Some students will opt to pay for their shirts a dollar or two at a time. You can also sell shirts to teachers, staff, and attendees. Order a few additional shirts to give to guests such as special speakers or your school superintendent.

When ordering shirts, a good spreadsheet is worth its weight in gold. Keep track of money collected for shirts on one page, with the size requested. We don't recommend distributing shirts before the day of your event, especially if using students as hosts. We've learned from experience that giving students shirts prior to an event means that someone forgets and leaves the shirt at home. If you are hosting multiple events, then you will have to trust your hosts to take home their shirts, wash them, and return wearing them when they are needed. You'll need to have additional shirts available if you sponsor multiple events in the same year, because you will recruit new hosts between each one. Be sure to pad your order with a few extras of various sizes to allow for errors. If you need to do a second run, the setup costs mean that you will pay more per shirt by running multiple orders than what you would pay for a single order.

Another way to dress up your volunteers is by the use of hats. We like to use hats or other headgear to express the themes for our events. Energy-themed events can make use of construction hats, ocean nights with a pirate theme can include bandannas, and biome events can use safari hats or visors. You might want to supplement your hosts' costumes with other decorations, such as leis for tropical-themed events or headbands with antennae for an Astronomy Night. Ordering these extras from a novelty supplier by the dozen can bring the cost down to a dollar or so each.

One final way you might accessorize your costumes is by using name tags with lanyards (see Figure 5.2). Name tags help attendees identify the volunteers who are hosting the event. They also build that sense of belonging and participation for your hosts. Laminate the name tags or use plastic badge holders so that they will make it through the event intact and can be reused. You can also use these to take attendance at any planning sessions. Place your name tags out on a table and have your volunteers or students move their own name tags from one location to another. You'll quickly know who isn't there so you can follow up with them later. If you are hosting multiple events at different locations, be sure to have a system of collecting name tags after each program. While not by any means necessary, we've found little touches like these can help your student hosts stand out from attendees and create a sense of belonging for your hosts.

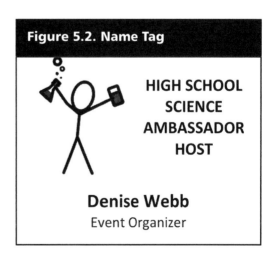

**Figure 5.2. Name Tag**

HIGH SCHOOL SCIENCE AMBASSADOR HOST

**Denise Webb**
Event Organizer

## Props

During Donna's first year of running the Astronomy Night program at her middle school event, she stumbled across inflatable aliens on sale at an online novelty store. These seemed a great way to add a bit of fun to the event, and Donna ordered several dozen. The aliens had an important role: One was posted at each classroom door where an activity was taking place and held a sign with the session title. Our aliens not only contributed to the atmosphere, but also provided lots of fun as hosts danced with them and incorporated them into their sessions. The aliens were stored and used again several years later when the astronomy theme was brought back again. We learned from this first year that inflatable mascots can make a program livelier, and we've incorporated them into all our themed events. (See Figure 5.3.)

**Figure 5.3. Donna With Principal Cindy Salloum and an Unidentified Alien Mascot in 2007**

Over the years, we have used a number of inflatable mascots: inflatable pirates for a nautical theme, dinosaurs for an energy theme (okay, that's a stretch, but why not?!), and monkeys for our Biome Safari (see Figure 5.4). We always opted for the largest size, usually 4–5 feet in height. Our student hosts always seemed to enjoy the fun environment these mascots created. Probably the most unusual mascots were selected for the year that the Discovery theme was used. The event program centered around famous discoverers and explorers in science, math, engineering, technology, and exploration. Sessions included an overview of each discoverer's contributions and a hands-on activity. We opted for inflatable mannequins and required that students dress them up in appropriate costumes for their discoverer's time period. This was probably one of the most unusual challenges we've given students. They responded with such creativity! Annie Jump Cannon was dressed in period clothing from a thrift store, Euclid was draped in a sheet, and Neil Armstrong was put in a white jump suit with a space helmet made from cardboard and aluminum foil. For the session on William Rankine, a physicist and engineer whose work paved the way for the modern roller coaster, students created a box for the mannequin to sit in that was decorated as a roller coaster car and suspended it over the door of the classroom.

Because of our ever-growing program, the mascots eventually became quite costly and if we wanted to continue having them, we needed a solution to their high costs. Eventually, we hit on an option that worked—mascot adoption! Early in the planning stage, we gave student hosts the opportunity to purchase, adopt, and take home a mascot after the event. Those who participated in the adoption program covered the cost of the mascot, named it, and received an adoption certificate. Using a permanent marker, the name of the mascot was printed on the front and "Adopted by … " on the back of each mascot. Certificates of adoption were checked at the exit to make sure no one took home an inflatable without the proper authority. The adoption program was a success in that it allowed us to enhance the festive atmosphere of our events for no cost.

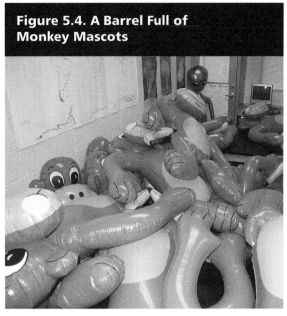

**Figure 5.4. A Barrel Full of Monkey Mascots**

Another type of prop you will need to add to your event is signage—and lots of it. Signs need to be posted inside and out to provide instructions and information. General signs can be branded with the artwork for the event or made generically for multiple years' events. Outdoor banners should be made of vinyl or plastic. We recommend placing an outdoor banner that says, "TONIGHT IS FAMILY SCIENCE NIGHT!" at the entrance to the school on the day of the event as a reminder.

Inside, you'll need signs on all the doors to your events. We have always tasked students with making these signs. Signs can be made from butcher paper for each activity, like the one shown in Figure 5.5 on page 58. A more reusable option is making door signs using shower curtain liners, permanent markers, and crayons. Shower curtain liners are inexpensive and make great canvases for student work. We like using these signs for our Science Ambassadors program since they can be folded up after each event, easily stored, and reused in future events. While regular markers tend to smudge on vinyl, permanent markers make bright, bold drawings. But they can get quite expensive. The best technique is to sketch on the liners in pencil, outline in permanent marker, then fill in using crayon.

**Figure 5.5. Door Sign**

## Sprucing Things Up

It is important to convey an atmosphere of fun and excitement in every corner of your event. The activities you select, how you "dress up" your volunteers, the extra activities you integrate, and even the decorations you choose should create an exciting atmosphere for your attendees. Decorating their activity spaces is one of the responsibilities that our student hosts have enjoyed the most. They use balloons, streamers, butcher paper, fabric, plastic, and more to dress up activity spaces, hallways, the cafeteria, and other common areas.

To get the student body of your school excited before your event, decorate a common area a week or so in advance. Create a three-dimensional exhibit to boost interest using your mascots, inflatables, or other props. On the night of the event, move your display to the entrance or allow your greeters to dress up the lobby on their own.

If you want your attendees to feel as if they've walked into an engaging atmosphere, decorating is more than just adding a few balloons and streamers. You want to create spaces that look, sound, feel, and smell exciting. Encourage creativity from floor to ceiling. Another year, students created a star map on Donna's classroom ceiling for a session about astronomy myths. Ceiling tiles were taken down and painted black, then constellations were created from glow-in-the-dark stars (please note that you must use flame-resistant paint). The students used silver markers to label the 59 constellations for reference. Figure 5.6 shows the results of that ceiling transformation.

To create a really special feel, students should attend to other senses, such as smell and sound. To provide the smell of a garden or the ocean, use a diffuser or incense. We don't recommend ever using candles, as they can create a fire hazard. Sounds are also important. Using music as attendees enter an activity or between sessions will help set the mood. With the right accents and props, you'll have created a festive atmosphere that engages your attendees in a fun and exciting event.

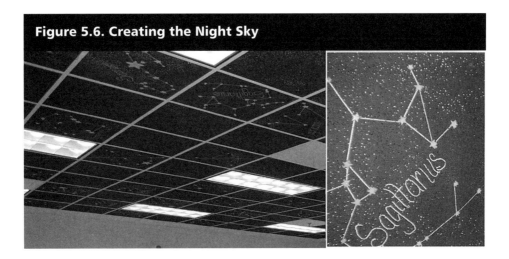

**Figure 5.6. Creating the Night Sky**

## Managing Your Costumes and Props

We can't overstate the importance of having a good management system to keep up with all the bits and pieces you incorporate into your event. Make a record of the supplies you have left over each year, and use those in future years for spicing up events. Now that you've worked out the costumes and props that you want for your event, we'll turn our attention to how to engage the community in your events.

# Chapter 6

# Inviting the Community

## Overview

In this chapter, we will discuss ways to integrate the community and its resources into your event.

- School Resources
- Community Resources
- Fundraising
- Getting the Word Out

Integrating other stakeholders into the Family Science Night event at your school can help build a culture of science that spans disciplines and departments and is the key to building a sustainable program. The more stakeholders are involved, the more support you receive and the greater the attendance. Inviting participation from nearby schools and the community means you can offer more activities for your attendees to participate in. There are multiple ways that your school and community can be involved. You might want to invite clubs and other organizations to have an active role in your event. You can also pull from community resources for guest speakers, for donations, or to sponsor activities. Regardless of how you broaden your event, the more involvement you have, the more successful it will be. In this chapter, we will present you with some ideas for reaching out and including school and community resources, describe how to share your program with the community, and discuss one of the most useful reasons to cast a wider net—getting donations to support your program.

# Chapter 6

## School Resources

There are many groups and organizations at your school that might want to take on a role at your event. Teachers in the arts can add to your event by integrating their content areas into science, while clubs and other organizations can also broaden the experience. Your art teacher can display student work in an art gallery, as ours did. A drama teacher might want to have students perform skits that tell the story of a famous scientist or discovery. The music teacher can include a recital in a common area. Technology teachers can also participate in the event by demonstrating educational or theme-based interactive websites. Student-created multimedia presentations can enhance the atmosphere in common areas. Be sure to ask teachers to participate early, especially if you are using a themed event, so that they can prepare and help suggest ways they can add to your event.

The Family Science Night event is a great time to collaborate with your media specialist by holding a book fair during your event. He or she will appreciate your attendees spending time browsing the latest paperbacks during an activity break. You might even be able to persuade your media specialist to highlight a special section of science or science fiction books. Another idea to broaden involvement is to suggest that your health or physical education teacher hold a health screening at your event. In addition, career education teachers can facilitate student demonstrations of everything from culinary science to dramatic performances.

Most schools have after-school STEM clubs such as science and math competition teams, and other clubs. Invite these groups to participate in your event by holding a demonstration, displaying their recruitment materials, or hosting an activity. Sometimes a specific class, department, or grade level will be involved in academic competitions, such as a science fair or math bowl. Let these groups take a place in the spotlight. Set aside a common area, such as a part of the cafeteria or the entrance hallway, for these groups to put up display boards that show off their accomplishments and achievements.

Interestingly enough, Denise's involvement with her first Family Science Night came out of a science fair competition held during her first year as an elementary STEM teacher. At one of the competitions, she put her students' projects on display in the lunchroom for parents to view and added some hands-on science activities at nearby tables. During this event, the tables were packed with all age groups, from preK to adult. She could barely get her guests to leave because they were having so much fun! For Denise, this demonstrated a need for providing this type of hands-on science interaction on a larger scale. The challenge became how to have more stations but not overburden her elementary teachers with already very full plates. Everything changed when she and Donna met and started the High School Science Ambassadors program, which allowed for participation in more hands-on science activities.

Former students returning to help run events has been one of the biggest surprises of our experience, and we've found that some keep coming back to help for years. Assigning them roles such as making announcements, monitoring halls, and assisting with ticket sales and surveys helps free up current students for providing more activities. In addition to inviting STEM-related clubs and organizations from the middle and high schools that your school feeds into, you might also consider asking those schools to send your former students to provide an orientation station for rising students and share information that next year's students will need to know. Encourage these schools and programs to provide information for your students about the opportunities that they will have for clubs and organizations in their new middle or high school. Former students who have moved up are perfect for this role!

We have six children and have been to more school functions that I can count: concerts, plays, sporting events, science and tech fairs, bingo nights, and book sales, as well as no less than 50 open houses over the years, to name a few. Out of all of these gatherings, Family Science Night was the only event that was hosted, staged, and executed by the students and had something to offer everyone that attended, parents and kids alike. It involved the entire school like no other event! The lessons outside of the scientific scope that the students learned were immeasurable and included hospitality, event planning, organization, project management, teamwork, and responsibility. I applaud and appreciate the pride and sense of accomplishment we saw develop in our daughter during her four years of involvement with Family Science Night (she went back her senior year of high school to help out).

—**Dawn Pruitt**
*Parent*

## Community Resources

Each community has its own flavor and a unique variety of science-related opportunities. Larger cities may have zoos, aquariums, science centers, museums, or performing arts centers that can bring a novel element to your program. Even smaller cities often have nature preserves or environmental centers. State parks located near smaller communities can offer unique resources that could be tied into your event. Businesses, especially STEM-oriented ones, may be able to enhance your program in multiple ways.

We have already discussed how Donna's middle school invited the local nature center to bring a portable planetarium to her events every year. She and the facilitator would discus the planned theme and ways to coordinate the planetarium show to match. This was always one of the most popular activities. For astronomy themes or events without a theme, just showing the stars that are visible and how to find them in the night sky is sufficient. However, you might need

> I value the partnerships and family connections that are an ongoing result of bringing everyone together to do AND understand science. The fun memories that are made benefit the public relations side of schools in the community. The experiences that are designed for a Science Night show science in a positive light for stakeholders who participate.
>
> —Dr. Tameka Osabutey-Aguedje
> *K–12 Science Specialist*

to get creative to tie this into other themed events. An Ocean Adventure program can focus on celestial navigation. This session also works for a biome theme, as celestial navigation has historically been important to traveling across the desert, as well as the seas. For the discovery-themed event, the focus was on astronomical discoveries.

There are dozens of other ways to include community-sponsored activities in your Family Science Night event. Having a "career fair" as part of your program is a great way to include the community. You might set up tables in the cafeteria for people in STEM fields to meet and greet your students or hold a brief demo or activity. Electricians could show students how to create simple circuits, veterinarians might bring a few animals, and doctors could set up a microscope with slides of different cells. If you are holding your event during the day, perhaps on a Saturday morning, you might consider including a "vehicle day" component. Invite local public service organizations to bring a fire truck, a police cruiser, an animal rescue vehicle, recycling trucks, and other vehicles to display.

Every community has local business and service organizations that can offer ways to enrich your event. Parks and recreation departments offer special programs or activities. Wildlife rescue could provide a birds of prey demonstration, and environmental organizations might offer other opportunities such as water quality testing. Businesses such as tech companies, medical labs, and landscapers can be contacted to see how they might participate in your event. The sheriff's office might be willing to set up a crime scene. One way to successfully involve these organizations is to reach out to the parents of your students and your school's partners in education. Let them know that you want them to participate in the fun! And, of course, fostering a good relationship with the community helps as you consider how to fund your program.

## Fundraising

Support from your school and local community is key to having a successful and well-funded event. Hosting a Family Science Night event is not without cost. Even if you are planning a very simple program that only includes hands-on activities, it requires money for supplies and materials. We've learned how to create some amazing experiments and demonstrations out of inexpensive materials, but providing materials for multiple activities for hundreds of attendees can add up quickly. We soon found ways to finance our Family Science Nights, all of them involving the school or local community at some level, and each year have discovered new sources of revenue. Had we known then what we know now, we could have had a better program with less stress in our first years. You will figure out what works to fund your events over time, but hopefully we can offer some suggestions to get your program started with better funding from year one!

## Before the Event

As soon as you know you want to hold a Family Science Night, it is time to start thinking about financing your event. Some funding options are easy enough. Your Parent-Teacher-Student Organization (PTSO), if you have one, may have funds set aside for just such a program. Title I programs often are required to hold a certain number of after school events and have funding to support them. Your state science teacher organization may offer minigrants, as ours does. In Donna's first year at the middle school level, the program was funded by a teaching award, in addition to a PTSO donation. Additionally, a number of stores and businesses offer grants to support their local communities. During our first year of running our joint Science Ambassadors program, we obtained a $1,000 grant from such an organization. Between this seed funding and other donations, we were able to cover the cost of the program. However, as our program grew, so did our financial needs. We began to be more creative and found ways to help build a better event.

Probably the easiest source of funds is to create a multipage program and sell advertising space (see Figure 6.1, p. 66, for an example of what a program might look like). We use landscape formatted pages, folded in half, to create small programs. We've used these multipage programs for events at all levels. For our Science Ambassadors program, a template is used for each activity, similar to those presented in Section 2, and a PDF file created to post online prior to the event for attendees to preview. Selling advertising space ended up much easier than expected. We send a letter out early in the year to parents, the school's partners in education, and nearby businesses. For our programs, a full page can bring in $250, a half-page ad costs $150, and a business card–size ad is sold for $75. Using ads, we have easily raised in excess of $1,000 each year to help fund our events. We also accept donations from businesses in exchange for advertising space. These donations can be used for materials or for a raffle. For example, a local grocery store that provides $150 of donated goods receives a half-page advertisement. Students can be given incentives to find sponsors, and a helpful parent might be willing to canvass businesses in the community. Local civic organizations, such as the Chamber of Commerce or the Rotary Club, might also provide leads for sponsorship. Keep track of your donations and be sure to send out thank-you letters or certificates to your sponsors, as they can use their donations for tax deductions. (More on thank you notes in Chapter 8.) We have learned that creating a culture of science in the school and community brings attention to your program, and advertisers will start to find you!

We found early on that our most overlooked resource was the parents of

**Figure 6.1. Program Cover**

our students. Sometimes you don't think about asking the most obvious source of help: parents. Once you know what materials and supplies you need, put out a wish list to the parents of your student hosts and to your PTSO. Be sure to get this list out as early as possible. If you are trying to collect 200 empty toilet paper rolls, it may take some time! Make use of parents' jobs, hobbies, and interests. The dad who sends you homemade cookies and candies for the holidays may be the right person to help with the bake sale. The mother who runs a gardening center might be able to donate seeds for one of your activities. The parent who works at a dentist office is likely to contribute the gloves you need for owl pellet dissection. Some parents are just happy to help. After sending home a wish list of general supplies, we find that most of the items we request are sent in, sometimes in an abundance far greater than what we need.

A number of opportunities can be used to raise funds throughout the year. Earlier, we mentioned that a Science Surplus Store can be a great extra resource to include at your event, but don't limit yourself to one night of sales only. Denise has had a great deal of success with her school science store during the school year. Spinning tops, materials left over from Family Science Night events, bouncing balls, magnets, tornado tubes, glow-in-the-dark plastic bugs, and poppers are some of her students' favorite items. Small items like these that can be purchased in bulk from online novelty suppliers are in high demand.

Donna has used an innovative strategy for fundraising throughout the year that was highly successful with her middle school students. Once a month, she hosted a science fiction movie night (Sci-Fi Fridays) in the school's media center. As with many things, these nights started out in her classroom with just her students but became more popular as her students asked to bring friends. If you want to use this activity as a fundraiser, you legally will need to get a permit for showing films. An annual license is a bargain, rather than paying a per-event fee, and can be used by multiple clubs and organizations within the school over the course of the year. Parents and students alike loved these opportunities. For us as teachers, these were great opportunities to interact outside the classroom in a way that built a positive rapport.

## At the Event

While you want your program to be free and open to the public, there are opportunities for raising financial support during your event. Family dinners, bake sales, and ticketed events, as discussed earlier, are all potential income producers to help offset the cost of materials and supplies. Your reception is a great place to sell raffle tickets for items donated by local merchants. One unique way Denise is now raising funds for the Family Science Night at her school is by using a "photo booth" at her event (see Figure 6.2). Attendees can get their pictures taken against a science lab backdrop, showing off the fun make-and-take science items they made in their favorite activities. Purchase these backdrops from an online novelty store, or have older students create different science-themed backdrops on butcher paper. Permanent markers and crayons on shower curtain liners make great, inexpensive canvases. Set up a room with a choice of backdrops, and ask for a dollar donation for students to get their picture taken in front of their favorite scene. Add a few science lab props, like beakers or a borrowed plasma ball from a local high school, to add some fun to the set!

You might consider positing a donation jar at the exit. Let attendees know in your program that you appreciate donations, and tell them where to find the donation jar at the end of the evening. You'll find that this allows you to start planning for future events with advance funding.

**Figure 6.2. Photo Booth Fundraiser**

## Getting the Word Out

Be sure to share the information about your program throughout your community. Make signs and banners to post at your school's entrance, and advertise on the school website and in newsletters. Create a video to share on your school news program or a display for your cafeteria to generate excitement. Invite your superintendent and school board members. Contact your local newspaper to cover the event, and put up student-created posters in local businesses. Eventually, your program will earn a reputation and the community will look forward to the event each year. And if you are taking your program on the road, you'll find families from one school will attend programs at other schools, as well as their own, to participate in more activities than they can complete in a single event! Holding Family Science Nights creates a culture of science in your school as well as your community.

# Chapter 7

# Showtime!

## Overview

In this chapter, we present the last-minute details and address challenges and obstacles that might arise on the night of your event.

- Before the Curtain Rises
- On the Stage
- After the Show

Family Science Night events are always exciting and present unexpected challenges. Nothing ever goes quite like you expect it to, but everything always works out. Right up to the day before your event, you'll find a dozen things to worry about and panic over. We always do! Some needed materials for an activity haven't arrived, or the flu is going around and several of your hosts are out, or the weather is rainy and the parking lot is flooded. Some of these things are in your control, and some aren't. So, fix what you can and work around what you can't. Over the years, we have found a number of strategies to help our Family Science Night events be successful, regardless of each impending catastrophe. In this chapter, we'll share how to handle last-minute items and what to watch for to make your event run smoothly.

### Before the Curtain Rises

On the day of your event, you'll find you have a number of last-minute preparations that need to be handled, no matter how well you planned. Hopefully, you've handled most potential problems prior to your event. Your hosts have practiced, your programs are ready, the materials have arrived, and you are ready to go. You have signs in place, you've confirmed your guests, and everything is in order. But there is still a long list of things to accomplish!

Final day preparations will vary according to where you are holding your event and who is running the activities. In Donna's middle school, the students running the event had an "in-school field trip" on the day of the event. They were excused from all their classes and spent the day doing last-minute preparations. They did a final run-through of their presentations and practiced their activities on each other. They created door banners for their sessions and decorated classrooms during each teacher's planning period (see Figure 7.1 for an example of door signs showing activity locations). They sorted their materials and organized them for efficient distribution. For the full day, students anxiously anticipated the event and worked hard with last-minute preparations. After school dismissal, students had a bit more time for final touches; then they grabbed a quick dinner before attendees arrived. It was always a true, authentic learning experience in which students assumed responsibility and accountability.

For our Science Ambassadors program, the day of the event looks quite different. Because they are not in the school that is hosting the event, there is little to no preparation they can do on the day of the event. They don't even know what room they will be assigned to or what it looks like until they arrive on site. But the anticipation is no less when they assemble at the end of the school day. Student hosts gather after school in one classroom, where they collect their materials and eat dinner (usually pizza delivered just after dismissal). Student hosts board a bus with their supply bins and travel to the elementary school where they will be holding the Family Science Night event. During dinner or during transit, they are given a map of the school with their assigned classrooms and general setup instructions. Once they arrive, they head to their assigned spaces to set up for the event. An additional room has usually been assigned as a "home base," where extra supplies are available and students can find help if they need it. There's usually quite a bit of rushing back and forth as hosts arrange furniture, put down table covers and drop cloths, find the nearest source of water or ice if necessary, and lay out materials.

Your volunteers, whether they are students or adults, are the most important aspect of an event for any manager to consider. If using students as hosts, you should already have planned with large enough groups to accommodate for absences due to illness, schedule conflicts, or lack of transportation. However, it is possible that one or two groups of students may be too small to effectively conduct their activity on the night of the event. If so, then you may need to shift people. We usually have four to five hosts that are given generic responsibilities or assigned as substitutes for each event. They are expected to be flexible and shift according to need. Sometimes we pull hosts from large groups and ask them to work with other activities where someone is absent. Occasionally, you may even have to shut down one activity. If there are enough other activities, then losing one to absenteeism will not be noticeable.

Supplies can sometimes present last-minute obstacles. You might find that important materials are delayed and don't arrive in time. Our Science Ambassadors have arrived for events and realized their supply box is back at our home school, or the student who has the supply bin may be sick. In these cases, you make do. If you are on the road, then chances are the building you are visiting has tape, scissors, pipe cleaners, rubber bands, card stock, twine, craft sticks, or whatever necessary material you are missing. Holding your first event each year at your school, or one very close by, provides a practice run for taking the program on the road, as it is easy to go back to retrieve missing supplies.

Weather can sometimes present unusual hurdles. If you want to conduct an activity outside, have a backup plan in place. If you want to do telescope viewing, great. But plan on an indoor "how telescopes work" session if it is cloudy. Launching film canister rockets outside is great in good weather but not a good idea in rain. Have a tarp ready to put down on the floor in the gym to reduce the mess. Bad weather can be a mixed blessing. Rain can mean canceled outdoor sports activities and an increase in your attendance, but severe weather can keep families from attending your event. For us, living in the South (where an inch of snow or a small amount of ice can shut down entire cities for days), we like to plan a backup date for events scheduled in winter months. Of course, if you do have to postpone, it is likely that there will be unanticipated schedule conflicts. So, be flexible and recognize that you may have fewer activities than originally scheduled.

**Figure 7.1. Final Preparations: Door Signs**

Of course, the most important things you must to address during the final hours leading up to your event are last-minute checks for safety and security concerns. Look for safety issues such as tripping hazards, and discuss responsibilities for children engaging in activities with your hosts to avoid injuries. Make sure each space you are using and the accommodations you have to make for the facilities meet your school's and district's safety regulations. You will need to use the building's public-address system during the event, so make sure you, and any student managers, know how to use it before your attendees arrive. Make a couple of practice announcements. Have and use two-way radios for communicating. Make sure your hosts know whom to contact if there is a problem and how to reach them. You should also review the building's evacuation plan in case of fire or severe weather with your volunteers. You may not think that you'll ever need to know it, but sometimes you do. One year during Donna's middle school event, the weather service issued a tornado warning, and a half an hour was spent with hosts, volunteers, families, teachers, and staff all huddled in the hallways. While we recommend emphasizing in advance that this is a family event and parents should not drop off children and leave, make sure that there are protocols in place for dealing with unattended children at your event. We suggest making sure that you have an adult stationed at the entrance to prevent parents from dropping off unsupervised children during the event.

Right before the doors open, take a final walk through all the areas that you will be using for activities. Make sure your activity hosts are ready, and scan for last-minute problems. Any preparations that are incomplete will most likely not be finished now, so ask your hosts to work with what they have, and get ready! It's showtime!

## On the Stage

Your major responsibility for the duration of the event is to make sure things run smoothly. First, you need to take care of your activity hosts, making sure they have what they need and helping them deal with any problems that arise. Second, you will need to manage any issues that come up with the attendees. However, event guests are seldom a problem if you are taking care of your activity hosts. When your volunteers are prepared and happy, they will be able to take care of the attendees. Your final responsibility is to make sure that everyone is safe and secure and that the building doesn't blow up or burn down in an experiment gone wrong. Just for the record, that isn't likely to happen! (See Figure 7.2 for a photo of guests arriving.)

Because event guests will start arriving early, you may decide that you want to gather your attendees in a common area until the actual starting time of the event. For our events, we hold guests in the reception area or cafeteria. When there is a pre-event meal, this is easy enough to do. If not, then have a plan for where they will wait. Maybe it's your book fair in the media center or the art show that has been set up in the lobby. Your activity hosts don't need families showing up while they are completing final preparations, so having a holding area is important. Before you start the event, you will need to make a couple of announcements to your attendees. First, remind them that the activity hosts are volunteers. If they are students, make sure you are clear that students will make mistakes and things won't be perfect, but that the student hosts have worked very hard and given their free time to prepare and present the event. Second, send

**Figure 7.2. Guests Arriving at the Reception Area**

different groups in different directions to begin the event. This prevents everyone from starting in the same place and reduces the demand on the activities closest to the front door. Spreading out the attendees to begin in different locations means that the activity hosts will be better able to accommodate them with less waiting and reduced stress. Over the course of your event, attendance at activities will still ebb and flow, but at least it won't start off with overcrowded activities.

Once the activities start, it will be busy, but in a euphoric way. If you've prepared well, everything will run like a well-oiled machine. Greeters will welcome attendees, hosts will run their activities, chaperones will be assisting guests, and attendees will be having fun. You aren't off the hook by any means, but at this point, what isn't done won't matter. Your attendees won't know what went wrong or what was supposed to happen that isn't happening. As the event manager, your job is to watch from the wings and deal with any major issues that arise.

You can't be in all places at all times, but you will need to circulate. Move through the building and check on your activity hosts to make sure they aren't having any problems. Watch for rooms that are overcrowded, attendees who seem lost or confused, and activity hosts who are having problems. When you see a session that is overcrowded, position yourself nearby in the hall and encourage attendees to go to another session to give activity hosts a chance to manage the guests they are currently working with. Just knowing you are checking with your hosts regularly helps them feel more confident and appreciated. If you have scheduled a longer event, two hours or more, be sure to give them a break and be sure to stagger the break times for different groups.

If you are running a session format for your program, you will need to make announcements on the PA system to start and end sessions. Make more than one announcement for the start of each new session, as the hallways can get noisy when crowded. Watch for any unsupervised children, and make sure attendees are not accessing any off-limits areas. Listen to the comments that attendees are making. Are there any indications of problems in an activity? Do you have hosts that are being exceptionally helpful?

If you are using students as hosts, remember your own advice—they are students and they will make mistakes. We've seen students who didn't know how to handle parents that were frustrated by the crowds. Occasionally, we've had to say something to activity hosts who were not doing their job or who were playing around. Although not a common occurrence, we've had to intercede with students who are not getting along with team members and move them to another role or group. On rare occasions, students will be less prepared than they thought they were and will struggle with their activity. In this case, you'll have to either make suggestions for fixing the issue or close that activity and ask students to work with another group for the event. No one wants student hosts to fail, but occasionally you have to allow learning experiences to occur, find a way to help them overcome their obstacles, and hope that growth is the result. We run dozens of activities a year and have never had more than one group each year of students that let us down, but it has been known to happen. For the most part, we find the experience of student hosts to be the most rewarding aspect of our program and have never had a problem with one of them that we couldn't handle.

Remember that during the event, you want a photographic record of the excitement and engagement of your attendees and volunteers. If you have a photographer designated for the evening, make sure that he or she is getting pictures of every activity and location (see Chapter 3). Be sure they know you want pictures of the people that are attending the event engaging in activities, not just shots of decorations and hosts. Have spare batteries around for the digital camera they are using.

As the event is winding down, you will want to make a few announcements. If using a flow model, give a 10-minute warning, and then make a second announcement 5 minutes before the end of the event. Once the event is over, regardless of which model you are using, make an announcement thanking guests for attending, and remind them that your activity hosts need to clean up before they go home, so they should make their way to the exit (see Figure 7.3

for an image of a student host cleaning up). If you are taking a survey, remind attendees where the drop box is. If you had a raffle, announce winners. You may even want to leave your coffee shop open a bit longer than the activities, especially if you have a fundraiser there such as a bake sale or book fair. Be sure to position yourself at the exit as people leave. Thank people for coming as they leave, and try to get a feel for what went right and what needs to change in future events.

**Figure 7.3. Student Host Cleaning Up After an Event**

## After the Show

Once the activities are shut down, there is very little left to do; the most important job is to clean up. You'll need to circulate to all the activity spaces used and assist your hosts with the process. They should know that they must return the room to its original condition or better but may need help making sure everything is reset. Do they need brooms? Mops? Paper towels? Did they forget trash or leave a mess on the floor? You don't want the teachers who shared their rooms to complain, and you want them to welcome you back again. Have your activity hosts collect and return any leftover materials and supplies. You will use these in future programs or to supplement supplies in your school's classrooms. Activity hosts may even need to bag up their trash, depending on the school's norms. Having a checklist prepared for your hosts with a materials list and general instructions can help with the process. Make a final walk-through to see that the building is cleared, everyone has left, and nothing is amiss. Check with the building's custodians before you leave to see if there is anything else to do.

Once you leave the building, make sure your volunteers and hosts have left or been picked up. If using students as activity hosts, then this is critically important for their safety. You may have to wait around for late parents or siblings to arrive. Have your cell phone charged and available for any calls they need to make. And remember that it is not their fault if their rides have not yet arrived.

You'll be exhausted after each and every event and you'll come away with ideas to improve for the next time. You'll find some things that don't work and some things that are more successful than you anticipated. You'll discover things that happen by accident that can be integrated on purpose. And if you are like us, you'll find this is the most rewarding activity you sponsor. We couldn't imagine our jobs without this annual experience, and we hope that what we've shared here will help you prepare awesome Family Science Night events for your school and community.

# Chapter 8

# Postproduction

## Overview

In this chapter, we'll look at postproduction tasks.

- Surveying the Audience
- Reflecting on Your Event
- Following Up With Stakeholders
- Taking Your Event on the Road
- Summary

The curtain has fallen and the last student has gone home. Is there anything left to do? First, get a good night's sleep. You've earned it. Then, it's time to review your Family Science Night event to evaluate what worked well, what didn't, and how to improve your program for next year. Perspectives from attendees, volunteers, and the administration can all inform your review process and help you plan for future events. You may even decide, as we did, that your program is such a success that you want to expand it.

### Surveying the Audience

We briefly mentioned in Chapter 3 that you might want to collect surveys from your attendees at the reception table. If you choose to survey your audience, be sure to keep it brief. Families that have spent their evening engaged in activities are often too worn out to give feedback. However, be sure that you have surveys available that they can either complete on the spot or fill out later and return with their children to school the next day. We've added surveys to the final page of our program (Figure 8.1). While you can use a Likert scale (a quantitative response with numbers ranging from 1 to 5 or 1 to 10), this type of rating is not overly useful unless you are gathering research. We found that asking a few short-answer questions will yield much more useful information to help you evaluate your event and prepare for future ones. Notice the questions asked in our sample survey shown in Figure 8.1. We asked for highlights and suggestions. We also like to ask a question about our student hosts to give them an opportunity to shine. Although it would be interesting to add questions about demographics, family size, or age of the students, we feel that it is more important to get information that will be useful in evaluation and future planning.

## Reflecting on Your Event

You will want to plan some time to debrief, but don't plan to debrief on the night of the event. Volunteers and hosts are going to be exhausted, and students will have homework to do. After our middle school events, we spent class time the next day talking about what went well and what didn't. With our High School Science Ambassadors, those conversations happen back at school as well, usually just prior to the next event or at an after-event celebration. If you are using teachers or adult volunteers only, you might plan a "thank-you" reception at the next faculty meeting or after class. Regardless of when you have this conversation, you will likely hear the same thing we hear year after year. We hear how much fun volunteering at the event was, how much everyone enjoyed it, how

### Figure 8.1. Attendee Survey

**Family Science Night Survey**

Please take a few minutes to tell us what you thought about this year's Family Science Night before you leave! We've provided pens and pencils at the reception area for you to make comments.

1. What did you think about this year's event?

2. What do you like best about our Family Science Night?

3. What could we do to improve the event next year?

4. What sessions would you like for us to recognize as having students who were well prepared and professional?

5. Is there anything else you'd like for us to know?

exhausted they were, and how they can't wait until the next one. It was just this type of feedback that led us to expand our high school program to host programs in multiple schools.

Some of the most interesting comments came from the middle school students who ran Donna's program for years. In every debriefing, with different group of students each year, Donna heard the same thing. Her students never knew teaching could be such hard work, they had a new respect for their teachers, and they didn't like it when people talked while they were trying to teach! One event would barely be done before student hosts started asking about the next year's program and possible activities to present.

During your debriefings, make sure you discuss what went well, what didn't work as well as expected, and possible improvements for the next year. Give volunteers the opportunity to provide input and recommendations. If your hosts are students and participation is part of a course, ask what they learned in terms of science content, leadership skills, and making presentations. Implement some metacognitive strategies, and ask them about the challenges they had to overcome and what they could do better. You'll want to have some large-group discussions, but allow for private reflection as well. Some students or hosts may not want to share their problems publicly. Be sure to take notes to refer to at the next event.

## Following Up With Stakeholders

To sustain your program over time, you'll want to make sure your community knows how grateful you are for its support. Providing a formal thank-you not only is in good taste but also will help your cause in future years. We create certificates for businesses that have donated to our program (see Figure 8.2) and have our student hosts write thank-you notes either in class the next day or at a later club meeting. Don't forget to produce certificates for your activity hosts. For students, those certificates can help document community service hours that many need for honors organizations. Once

**Figure 8.2. Donor Appreciation Certificate**

*In Appreciation to:*
Business Name
*Dedicated Business Partner of the*
*High School Science Ambassadors Program*
*You Make a Difference!*

*Denise Webb*
*Program Director*

*Donna Governor*
*Program Director*

your program has been in place for a few years, you will find it even easier to find sponsors. Remember to thank the parents, teachers, custodians, and other volunteers who have given their time to make your Family Science Night event a success. Send out a thank-you e-mail the next day that reinforces the importance of the support that you depend on for your program. Over time, these little gestures can help your program grow.

In addition to sending out acknowledgments and notes of appreciation, you should also follow up with the administration at the school where your event was held. Inquire about any specific issues that happened that you might not be aware of. Check to see that all areas were cleaned up, ask about what they are hearing from the families that attended, and find out what feedback they can offer. Encourage them to make suggestions for future years. If you held the Family Science Night at another school (more on this in the next section), discuss the event with your contact to find out what went well and any changes that should be made in future programs. Take notes, because it will be several months before you start collaborating on the next year's event. You'll be glad you did.

## Taking Your Event on the Road

Now that your first Family Science Night event is behind you, you might consider taking your program on the road. For teachers providing a single one-night event in your own school, you can skip this section. However, if you are using older students who have enjoyed sponsoring a Family Science Night event at an elementary school, you may find that one event isn't enough and your students are hungry for science adventures. Our High School Science Ambassadors program provided events for as many as six elementary schools in one year.

## Checklist 8.1.

### Hosting School's Responsibilities

- ☐ Obtain permission from the administration
- ☐ Set a date that will work for both your school and the visiting hosts
- ☐ Arrange for extra personnel such as school resource officers and custodians
- ☐ Advertise your event to your school
- ☐ Coordinate with the program manager to assign spaces and create a map for the program
- ☐ Conduct fundraising to contribute to the cost of the program
- ☐ Plan for extra activities to be included (e.g., meals, bake sale, career fair)
- ☐ Arrange for a reception area with greeters
- ☐ Make sure your attendees understand that the hosts are guests and that the attendees should be patient with them
- ☐ Provide radios for communication
- ☐ Make available necessary equipment such as ice machines, refrigerators, and cleaning supplies
- ☐ Help clear the building at the end of the event
- ☐ Send thank-you notes to your guest hosts

While we found that a bit overwhelming and trimmed down to four programs a year for sustainability, our Family Science Night is so popular that we have had to turn down requests from more schools than we can actually provide events for. (It has been our hope that eventually other elementary and high schools in our district would pair up and provide similar programs for their communities to expand the program to other schools.)

If you are planning to take Family Science Nights to other schools, then you will have to consider logistics for materials, coordinate with the other schools, and make a plan to pay for the additional cost. Materials can be managed by using plastic bins for each activity. We purchase our bins early in the planning process, and as materials arrive, we sort them into the proper bins. We label each bin and tape a supply list to the inside of the lid.

**Figure 8.3. Sample Door Sign**

We try to stock each bin with materials for at least 200 students, which represents the average number who circulate through each activity in a single event. Our student hosts transport their bin to and from each event and provide us with a restocking list after each one. We don't schedule our events any closer together than every two weeks to allow for time to recover and reorder supplies (or reschedule due to inclement weather). In the bin, students also store their name tags and the door signs they have made from shower curtains (Figure 8.3).

At the end of our last Family Science Night, activity materials are not restocked because of the uncertainty about which activities will be repeated the next year. However, because everything is organized in bins, it makes for easy storage from year to year. To fund materials for multiple schools, we charge each school a basic fee to partially cover the expense. To keep the program affordable, the fee doesn't nearly cover the full cost of the program. So, if you provide events for more than one school, you will need a bigger budget. We offset the cost with program advertising (discussed in Chapter 6), which is even easier to obtain when your program provides Family Science Night events at multiple schools. Advertisers can come from a broader geographic area, and the lure of a

**Figure 8.4. Denise and School Contact Amber Hoke at Amber's Family Science Night Event**

larger potential reach is attractive to more businesses. You may find other ways to raise funds work better in your community, but for us, advertising has allowed us to support our expanded program.

Taking your Family Science Night program on the road allows you to support and give back to your community. This is best managed through a formal program, such as our High School Science Ambassadors, where older students are trained to be the activity hosts for these events. This type of program requires a working relationship between teachers at the elementary and secondary levels, like the relationship we have. Denise handles logistics at the elementary school, while the high school students that Donna manages become activity hosts. However, once we started taking these events to other schools, we found another level of coordination was required. Denise and Donna worked in unison as our program's sponsors, while coordinating with a contact at the elementary school (see Figure 8.4). Our job was to manage and transport the student hosts, determine activities, provide the materials and supplies, create the program, and communicate our needs to the schools we traveled to.

For each school, you need a specific contact who will work closely with you on your program. It will be his or her job to advertise the program, schedule additional activities to include in the event beyond the ones we provided (such as a school dinner), and arrange for supervisors, monitors, and custodians. Your contact person will need to work with you on scheduling the spaces for your activities and providing a map to insert into the program. The contact should also be willing to reach out to partners in education

I love getting to work with the little kids and watching their faces light up as they learn and participate in all the experiments! I'm glad I was able to be a part of the Science Ambassadors club.

—**Hannah Daniel**
*Student*

to seek out potential program advertisements. There are additional tasks that need to be attended to, the most important ones shared in Checklists 8.1 (p. 80) and 8.2. Our elementary school contacts have worked with us for several years now and appreciate having an activity-based science program brought to their school. Amber Hoke, the science specials teacher at one of the schools our program serves, speaks highly of the program: "Having the Science Ambassadors host our Family Science Night offers two big perks: It gives our teachers and me the gift of time because the activities and materials are planned and purchased, and we can enjoy the event with our families. Also, it allows our elementary students to see high school students having fun with science and engineering—that brings the cool factor."

If you have decided to take your Family Science Night events to other schools, we have a few additional tips to help you keep your sanity. Schedule your first event each year at your own or a close-by facility. Your activity hosts will forget things and you will find you have needs you didn't expect. If you are at, or very close to, your home school, you'll find it easy to run back for what you need. Make sure your program includes dates for all your locations. There have been a number of parents who take their children to multiple events because they can't do all the activities in one night or go to another school's event because of personal conflicts on their school's Family Science Night. Let attendees know they are welcome to attend any of the scheduled events. We found taking our programs to other schools to be one of the most rewarding aspects of working with our Science Ambassadors, especially for our high school students, who seemed to enjoy every aspect of the program. With multiple opportunities to

## Checklist 8.2.
### Organizer's Responsibilities

- ☐ Set dates with contact at other sites (plan weather backup dates as well)
- ☐ Plan the core activities and sessions
- ☐ Recruit hosts and obtain parental permission if working with students
- ☐ Assign sessions to host groups
- ☐ Provide planning and practice sessions for your hosts
- ☐ Coordinate with other school programs to make sure that students will not have a conflict on the event nights
- ☐ Work with your contact to arrange activity spaces; communicate any special requirements for specific activities
- ☐ Assemble and print programs that can be used at multiple locations
- ☐ Arrange for shirts and name tags for hosts
- ☐ Feed your hosts prior to the event
- ☐ Transport your hosts to the site
- ☐ Gather, store, and transport materials and supplies—replenish as necessary
- ☐ Make sure that your hosts clean up at the end of the event and are picked up by their parents
- ☐ Arrange for email and group communication (possibly using a group app) for sharing information

share hands-on science with elementary-age children, these students were able to improve their activities at each iteration and felt a level of accomplishment that a single event night wouldn't have allowed. In all honesty, our part as program coordinators wasn't too much more difficult than if we had done only a single event. The majority of the work is in the preparation, and once the first Family Science Night event was done, it was relatively easy to restock and take the program to another school.

## Summary

Watching young children engage in hands-on activities and get excited about science is a continual joy. Having older students host events is the most rewarding aspect, in both the middle school session model and our High School Science Ambassadors program. Because of their leadership role, they learned science at a deeper level, developed leadership skills, and became role models for our young attendees at all our events. Seeing the enthusiasm and excitement of our student hosts, receiving feedback from the community, getting requests from additional schools to share our program, and having the support of the schools and teachers we have reached every year now for more than a decade, we have become firm believers in the importance of Family Science Night events.

# Section 2

## On the Stage

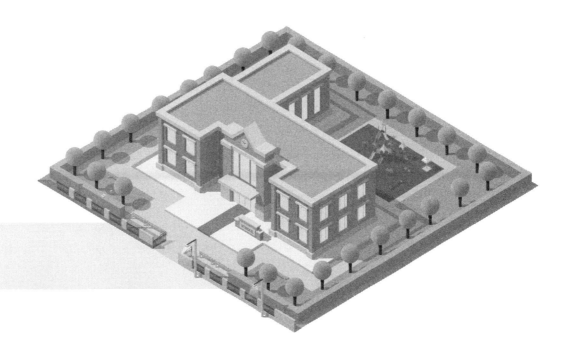

# Chapter 9

# Introducing the Activities

## Overview

In this chapter, we will introduce Section 2 of the book and the format used to present ideas for Family Science Night activities.

- Activity Template
- Using the Activities

At any of our Family Science Night events, the energy is electric. Even though we work hard to create a festive atmosphere for our attendees that they notice from the moment they arrive, you'll find the most excitement where the activities are taking place. Regardless of whether the program uses a session or flow model, if you wander into an activity station, you'll find parents and children actively engaged in hands-on science. Families move quickly from one activity station to another, trying to get in as many activities as possible. Laughter, excitement, and the language of science can be heard in every room. And whether families are launching rockets, exploring magnetism, or unearthing fossils, it is in these activities that the real magic of a Family Science Night event occurs (see Figure 9.1, p. 88).

So far, we've shared the key elements of how to plan and run a Family Science Night program, including tasks, materials management, programming, community involvement, creating a festive atmosphere, and fundraising. But we have yet to discuss the most important aspect of a Family Science Night—the activities.

# Chapter 9

**Figure 9.1. Students Exploring Potential and Kinetic Energy at the Hall Rollers Activity Station**

The activities are the core of any family science program. Over the years, many of our hosts have found activities on their own. One of our favorites was a session on steam engines, where attendees made "steamboats" from foil miniloaf pans, tea candles, and copper tubing. However, sometimes hosts need more help in finding an activity, so, more often than not, we provide a choice of activities along with suggestions for how to adapt our activities to meet their interests.

We've found that whether using teachers or students, you do need to make sure your hosts understand the science behind their activities. You may need to explain to your activity hosts how a circuit works, what potential energy is, or that sound is caused by vibrations. For our student hosts, learning the science is one of the expectations, and it is necessary to make sure that they won't foster misconceptions when working with young children and their families. Also, student hosts will not necessarily know the right questions to ask children as they work through an activity. So, we've learned to provide our activity hosts with some basic information about the science behind the activity. As for the attendees, we found that families are often frustrated that they can't complete all the activities at one of our events. So, we like to provide some information about our activities for parents to engage with their children in hands-on science activities they couldn't get to during an event later at home. Over time, we've compiled an abundance of activities with information that can be used to help hosts and parents. The following chapters of this book contain dozens of the activities we've used at our Family Science Night events.

## Activity Template

To make these activities more useful, each activity is presented in a two-page layout. A sample activity is included here (pp. 90–91) to guide our discussion of the layout. The first page is a "Teacher Tips" page that includes elements of three-dimensional teaching. The *Next Generation Science Standards (NGSS)*, and many recently revised state standards, are research based and implement a three-dimensional approach to learning. This refers to instruction that integrates science content, science and engineering practices, and crosscutting concepts. It is a research-based focus that purposefully engages students in learning science by participating in the practices of science and integrating cognitive tools (crosscutting concepts) to develop an understanding of science concepts (NRC 2013). The template used to present activities in this book specifically addresses those elements of three-dimensional learning for each activity.

> This will be the fourth year for our Family Science Night, and each year the attendance grows. It is the most fun and successful free family event we have all year as it is designed specifically for hands-on discovery learning by kids and for kids. The high school students are the facilitators and are amazing teachers to all grade levels. The students run from one station to the next learning and discovering about all aspects of science, technology, and engineering. This night is fun for the entire family and one our families look forward to each year.
>
> **—Barbara Vella**
> *Principal*

Also included on the Teacher Tips page is specific information about how to manage the activity, including supplies and materials, in "Before," "During," and "After" the Event sections. We've also added an "Additional Resources" section to include items that will help build supporting stations to supplement the activity. When possible, we suggest including some of these resources and activities nearby to give attendees something extra to do. These additional touches provide a richer experience as well as help keep attendees occupied when hosts are busy. The final section included on the Teacher Tips page is called "Things to Think About."

| SAMPLE—Teacher Tips: WORK IT, CIRCUITS! | |
|---|---|
| Core Content | Conservation of Energy and Energy Transfer |
| Crosscutting Concepts | Energy and Matter |
| Science and Engineering Practices | Planning and Carrying Out Investigations |

| | |
|---|---|
| Guiding Question | How does a circuit work? |
| BEFORE the Event | Old Christmas light strands work well for this activity, as well as for a student take-away. Ask parents for old unused light strings, and look at garage sales, thrift stores, and after-Christmas sales! Include some additional activities for students to explore at the station on their own, such as a basic circuits kit and energy balls. Discuss the need for eye protection. |
| DURING the Event | Use AA batteries. DO NOT directly connect the wires or clips on each side of the battery without the light bulb. Without the load (the light bulb) the wire can heat up quickly and can burn fingers. |
| AFTER the Event | Leaving batteries connected runs down the battery quickly, so be sure to store your equipment with the batteries in a separate bag or storage bin. |
| Additional Resources | Provide an energy ball and a circuits kit for students to explore further. |
| Things to Think About | 1. How does energy get from the battery to the light bulb? (It flows through the wire.)<br>2. How can you make the light work? (There are several correct ways to make a circuit. The wire must connect both the positive and negative terminals of the battery with the bulb along the path of electricity. The wire must touch both the bottom and side of the metal part of the bulb to work.)<br>3. What will happen if you put TWO or THREE light bulbs in the circuit? (This is a series circuit. Bulbs will be dimmer.)<br>4. What do you think you could do to make the light bulbs brighter? (One way to accomplish this is by putting another battery in the circuit. You can also guide students to build a parallel circuit with several pieces of wire. The bulbs will remain bright and not dim.) |

## SAMPLE—Activity: WORK IT, CIRCUITS!

### Program Information Section
*List location and hosts' names*

| | |
|---|---|
| **Science Behind the Activity** | Circuits are pathways that electrons flow through. A battery or other power source provides the electrons that move through the wires. As electrons move through the circuit they can do work—for example, make a light bulb light up. For electricity to flow, the pathway must make a complete circuit and return to the source. Circuits can be constructed in series or parallel. Series circuits have a single path that flows through more than one device, while parallel circuits allow current to flow through multiple paths. |

Simple Circuit

| | |
|---|---|
| **Vocabulary** | • Current—the flow of electricity through a conductor<br>• Circuit—the path in which the electricity can follow; can be closed or open<br>• Closed circuit—has a complete path that allows the current to flow<br>• Open circuit—has a break in the path so the current cannot flow |
| **Safety** | Generally, there are no concerns using low-voltage batteries. Safety glasses or goggles are required for this lab. Use caution working with wires, as they can cut or puncture skin. Bulbs can shatter and cut skin. Be sure to tell students to NEVER experiment with electricity from wall outlets. |

**What You Need:**

☐ Light bulbs from Christmas lights with wire ends exposed
☐ Circuit ball
☐ Extra wire sections with the ends exposed
☐ Batteries (AA only)
☐ Safety goggles

**How-to:**

1. Make a complete circuit by using a light bulb and a battery.
2. Make a series circuit by connecting more than one light bulb in a row (twist the ends of the wires together).
3. Make a parallel circuit by twisting the wires from two light bulbs together with a third wire on one side and a fourth wire on the other. Then, attach the new wires to the battery.

**Things to Think About:**

1. How does energy get from the battery to the light bulb?
2. How can you make the light work?
3. What will happen if you put TWO or THREE light bulbs in the circuit?
4. What do you think you could do to make the light bulbs brighter?

**Instructions and General Information**

### Estimated Activity Time: 10+ minutes

These are the leading questions that your hosts should be asking children as they participate in the activity. These questions match the ones on the activity sheet; however, sample answers are provided here for your hosts. It would be a good idea to discuss these questions with them prior to the event, and elaborate on any of the concepts that they may be struggling with before they engage with young children.

The activity page is formatted so that it can be easily inserted into your event program. (We recommend a landscape format with two per page.) In our experience, families that were not able to get to all the stations wanted to do additional activities with their children at home. So, for events using the flow model, we designed a page that could easily fit into our program. We felt it was important that this page not only address the activity and how to complete it, but also include some information on the basic science behind the activity to help explain the concept. Essential vocabulary is included to help develop the language of science. Safety considerations are also specified; however, we are very careful to include activities that pose minimal risk. Most of our materials and supplies for activities held at our events are common grocery store items, not only for safety reasons but also for easy accessibility and cost considerations. Of course, it makes sense to include lists for materials and instructions, but on this page, we've also included our "Things to Think About" questions. These are the guiding questions that hosts should be asking children as they complete the activity. These questions get at the important concepts children should be thinking about as they build conceptual understanding. Our questions are tied to the elements of three-dimensional learning highlighted in the activity. The questions correspond to those on the Teacher Tips page, where a basic response is provided. We've also tried to include, when appropriate, pictures or illustrations to help you envision the activity.

One important piece of information included on these activity pages is a time estimate for completing the activity. These are based on what we've observed with the families that come through our Family Science Night events. If the activity page indicates the time estimate is 10 minutes, that means that the families that pass through that activity station will generally spend about 10 minutes engaged in the activity. They make take longer, especially if some additional activities, such as those in the "Additional Resources" section of the Teacher Tips page, have been set up. Of course, if parents are doing this activity at home with a student, they will need setup time as well, but this is not something we've factored into our time estimates. Using a session model, the activity limit will be based purely on the length of your event, and hosts will likely include a group presentation, a demonstration, and a discussion or sharing activity in each session. With a 90-minute flow model event, most families will only be able to get to six to eight 10-minute activities, allowing for travel time between stations. Some will spend less time at a station they don't find as interesting to squeeze in more activities. So, plan accordingly and expect that the time your families spend in each activity session may vary when using a flow model (see Figure 9.2).

**Figure 9.2. Students in the Science Ambassadors Program Preparing Their Activity**

## Using the Activities

The activities we provide are tested at the K–8 level. However, we've decided to present them in three different chapters, based on the conceptual level. We've divided the activities into three levels, novice, intermediate and advanced. We do not recommend tying these concepts to grade levels, as we've found from experience that the same activity can be used with children of different ages, depending on the students' prior experiences, level of conceptual understanding, and interest. Many older students need novice-level experiences that help them build basic science concepts, and many younger children are more than capable of grasping advanced ideas presented in advanced activities if experiences are provided in an appropriate environment to construct meaning. However, we have divided up our activities into these three groups to help you select a range of activities at different levels for your Family Science Night event. Activities at the novice level include primary-level concepts that require minimal fine motor skills and can be completed in a shorter time period. Intermediate-level activities include concepts that are more complex, often engage learners longer and require better fine motor skills. Finally, activities at the advanced level are at higher conceptual levels, require more advanced fine motor skills, and will generally require more time to complete. When younger children are exploring more advanced activities, reframe questions using more basic language and build the experience for later conceptual exploration. For older learners working with lower-level activities, adjust the questions to stimulate more in-depth thinking and help address any misconceptions that might be present. It is our recommendation that you include activities from all levels at your events to accommodate a wide range of learners. You might also consider putting activities with similar concepts but at different levels together in the same space.

Activities presented in Chapter 10 are those we consider to be novice level, while Chapter 11's activities are intermediate, and those in Chapter 12 are more advanced.

We have large crowds at Science Night at our school. Children look forward to it because of the interactions and play. Parents are just as engaged! Learning together as a family has value, and there is a lot of discussion in multiple languages. The past connections of high school students who went to our school coming back as Science Ambassadors bring a lot of pride. I see the high school students teaching and think about how far they have come as students. Everyone benefits from participating.

—**Polly Tennies**
*Principal*

Start small your first year. Choose a dozen or so different activities that you can easily implement. Each year, you can grow your program by adding more activities and extra dimensions as shared throughout the book or implement some original ideas of your own. Change out the activities each year so that regular attendees can look forward to something different. Keep a few of the most popular activities, however, to give continuity to your event and for those attendees who heard about them the year before. Keep in mind that these activities can also be used for themed events. Get creative about how to make connections. For example, Foil Boats can be used with an oceans theme to focus on buoyancy, a biomes theme for a trip down a river habitat, or a discovery theme for a look at famous explorers from the Age of Discovery.

As you explore the activities in this book, let your imagination see ways to not only include but also polish and embellish the ideas we've presented. Think about ways to make each activity interactive and part of an exciting environment. Find ways to add self-guided stations to the activity space, and decorate to create an inviting atmosphere. There are endless ways to build your Family Science Night event. We hope that this book and the ideas and activities included here are just a start for you. May your program helps create a new generation of students of students who see science as something to participate in that is exciting and fun. Start small, think big! Before you know it, you will have built an event that builds a culture of science in your community!

## Reference

National Research Council (NRC). 2012. *A framework for K–12 science education: Practices, crosscutting concepts, and core ideas.* Washington, DC: National Academies Press.

# Chapter 10

# Novice-Level Activities

The activities presented in this chapter are what we are calling novice level. They either are tied to disciplinary content ideas that would be at a primary level or are simple ideas that require minimal fine motor skills. Don't limit these activities to using them only with younger learners. It has been our experience that older students often need basic experiences that will help them understand basic ideas in science. Novice-level activities such as these can address misconceptions and help build a stronger foundation for truly understanding more complicated concepts in older learners. For older students, ask different questions as they complete these activities to develop more advanced concepts. When choosing activities, we recommend you balance the number and type of activities at your event by both level and concepts. The activities here are as follows:

- Balancing Bugs—work with symmetry to balance a "bug" on your finger
- Bubble Olympics—explore surface tension by making bubbles
- Catapults—explore forces and motion in a simple engineering activity
- Catch the Wave—put an ocean of waves in a bottle
- Changing-Color Slime—mix primary colors of slime to make secondary colors
- Foil Boats—build a boat that holds the most weight in this engineering challenge
- Getting Buggy—explore the world of camouflage with plastic insects
- Grassy Pets—discover what plants need to grow
- Harmony Harmonicas—make sounds with a homemade instrument
- Ice Cream—explore the states of matter with everyone's favorite
- Maracas—make your own percussion instrument from plastic eggs
- Pinwheels—watch as wind energy moves your pinwheel

| Teacher Tips: BALANCING BUGS | |
|---|---|
| Core Content | Conservation of Energy and Energy Transfer |
| Crosscutting Concepts | Structure and Function |
| Science and Engineering Practices | Developing and Using Models |

| | |
|---|---|
| **Guiding Question** | How can you balance a bug on your finger? |
| **BEFORE the Event** | Cover tables where the students will be coloring. Have a variety of weights available in cups (pennies, large and small paper clips). Have several bugs precut, including some to use as templates to trace. |
| **DURING the Event** | Students will be making symmetrical bugs to balance. They can design their own on folded index cards or can use a precut shape as a template. If students choose to draw their own bugs, make sure the cards are folded first so they will be symmetrical. Also, make sure they include a point at the head, for balancing on the students' finger. Encourage students to try to design their own bugs, but if their fine motor skills are not developed enough, they may need to trace a bug or use a precut shape. The "wings" of the bug must extend forward beyond the balancing point to have the weights balance the "tail." |
| **AFTER the Event** | Make sure area is cleaned up and no marks are left on tables from markers. |
| **Additional Resources** | Have other balancing toys for students to explore while you are working with other students. |
| **Things to Think About** | 1. Why doesn't your bug fall off your finger? (It is balanced because it is symmetrical—it is the same on both sides and front to back. The balancing point is at the center of gravity.)<br>2. Why is balance important? (It keeps things upright.)<br>3. What are some examples of people using balance? (Ice skating, riding a bike, gymnastics.)<br>4. What are some examples of animals using balance? (Many animals use their tails to help them balance.) |

## Activity: BALANCING BUGS

### Program Information Section
*List location and hosts' names*

| Science Behind the Activity | Gravity is the force that pulls two objects together. Weight is the force of gravity on an object. The amount of weight tells us how hard the force of gravity is pulling us toward Earth.<br><br>The balancing point is where the center of gravity is. This is where the mass of the object is evenly divided and where the push and pull forces are working evenly. Any object can be balanced at its center of gravity, but it is easier to identify the center of gravity with objects that are symmetrical. |
|---|---|
| Vocabulary | • Gravity—the force that pulls two objects together<br>• Weight—the force of gravity on an object<br>• Mass—the amount of matter in an object<br>• Center of gravity—the balancing point<br>• Symmetrical balance—when all the elements on both sides of the center are equal |
| Safety | Provide scissors that are appropriate for a variety of ages. |

**What You Need:**

☐ Index cards

☐ Scissors

☐ Markers

☐ Tape

☐ Weights (pennies, large and small paper clips)

**How-to:**

1. Choose a bug shape or draw one on an index card.

2. Color both sides symmetrically.

3. Cut your shape carefully.

4. Put weights on the front wings of your bug to balance the mass.

5. Balance your bug on your finger!

**Things to Think About:**

1. Why doesn't your bug fall off your finger?

2. Why is balance important?

3. What are some examples of people using balance?

4. What are some examples of animals using balance?

**Instructions and General Information**

**Estimated Activity Time:** 5+ minutes

| Teacher Tips: BUBBLE OLYMPICS | |
|---|---|
| Core Content | Structure and Properties of Matter |
| Crosscutting Concepts | Structure and Function |
| Science and Engineering Practices | Planning and Carrying Out Investigations |

| | |
|---|---|
| **Guiding Questions** | What soap makes the best bubbles?<br>Can you make bubbles in different shapes? |
| **BEFORE the Event** | Cover tables with tablecloths. You should have standard-sized bubble wands for Station 1. You can purchase them, make these, or scavenge them from purchased bubbles.<br><br>Station 1: Have your tubs of bubble solutions set up in front of the bottles of different kinds of soap used. Provide a tub of clean water to wash the wands between tests. Use standard-sized wands at this station.<br><br>Station 2: Paper plates with a waxy surface will give you better results. Prepare the bubble solution. Glycerin will make bubbles stronger. It is helpful to put solution in squeeze bottles.<br><br>Station 3: Make wire bubble wands of different shapes for experimenting (e.g., circle, square, triangle). Have pipe cleaners to create different wands. Have three-dimensional "bubble wands" (e.g., cube, pyramid) to show how different surface shapes can be created. |
| **DURING the Event** | As bubbles pop and students experiment, soap will end up on the floor. Watch to make sure that the floor doesn't get too slippery. Have towels or paper towels ready to clean up spills. Provide plastic sandwich bags to put the bubble wands in for students to take home. Have a bottle of water ready or access to an eye wash station in case anyone gets soap in his or her eyes. |
| **AFTER the Event** | Make sure all soap is cleaned off tables and floor. |
| **Additional Resources** | If you have a projection board or computer, show slow-motion videos on bubbles popping. You may want to provide students with small, novelty-sized bubble bottles to take with them. You might also add a station where students use a pipette to put drops of plain water, and drops of water with soap on a penny. |
| **Things to Think About** | 1. Which solution gave you the biggest bubble? (Answers vary, depending on detergents used. Adding glycerin will make a bubble bigger and last longer.) Which solution made the longest-lasting bubble? (Same as above.)<br>2. Were you able to make a bubble of a different shape? Why or why not? (A simple free-floating bubble will only make round [spherical] shapes because of surface tension. It is the shape that allows the most air with the least amount of "skin." However, a bubble formed within a shape always finds the "shortest path" to connect the shape!)<br>3. What makes bubbles pop? (The "skin" evaporates and the bubble pops!) |

| Activity: BUBBLE OLYMPICS |
| --- |

**Program Information Section**
*List location and hosts' names*

| Science Behind the Activity | A bubble is air wrapped in a "skin" of soapy water. The "skin" is a thin layer of water between two soapy layers. Surface tension holds the bubble together, like molecules holding on to each other. If a bubble is stretched too far as you are blowing it, it will pop. Once the bubble is released, the surface of the bubble begins to evaporate, which will also cause it to pop. Glycerin can make the bubble skin thicker so that it takes longer to evaporate. On a cold day, bubbles will go higher because the air from your lungs is warmer than the air outside, and it will make the warm bubble rise. | Soap →<br>Water →<br>Soap → AIR |
| --- | --- | --- |
| Vocabulary | • Surface tension—the attractive force that pulls molecules of a liquid into the tightest possible shape<br>• Elasticity—how far something can stretch<br>• Evaporate—when molecules of water escape from a liquid | |
| Safety | Bubble solution is very slippery. Everyone should have a fresh new straw. Warn students to not drink bubble solution when blowing with straws. Make sure all students throw away their straws when done. | |

**What You Need:**

Station 1:
☐ Bubble wands
☐ Variety of dish soaps
☐ Glycerin
☐ Timer (stopwatches)

Station 2:
☐ Paper plates
☐ Drinking straws
☐ Bubble solution (with added glycerin)

Station 3:
☐ Bubble solution with added glycerin
☐ Premade wands of different shapes
☐ Pipe cleaners

**How-to:**

Station 1: Dip a bubble wand into one of the bubble solutions. Gently blow a bubble. Rinse your wand and try different solution. Compare your bubble sizes. Use the timer to find out which bubbles last the longest.

Station 2: Put some drops of bubble solution onto your plate. Put one end of the straw on the plate. Gently blow your bubble.

Station 3: Try to blow bubbles of different shapes. Make a wand!

**Things to Think About:**

1. Which solution gave you the biggest bubble? Which solution made the longest-lasting bubble?

2. Were you able to make a bubble of a different shape? Why or why not?

3. What makes bubbles pop?

**Instructions and General Information**

**Estimated Activity Time:** 10+ minutes

| Teacher Tips: CATAPULTS | |
|---|---|
| Core Content | Conservation of Energy and Energy Transfer |
| Crosscutting Concepts | Cause and Effect |
| Science and Engineering Practices | Constructing Explanations and Designing Solutions |

| | |
|---|---|
| **Guiding Question** | How can you design a catapult that will launch a marshmallow? |
| **BEFORE the Event** | Make sure you purchase plastic cups that are strong enough to hold the rubber bands without collapsing. You may want to test out several types. You can sometimes get your mini spoons donated by a local ice cream shop (they are the taste tester spoons). Set up a testing area with cones and a target. |
| **DURING the Event** | Students will want to try over and over. To keep the lines moving, allow two tries, then send them back to the table to redesign before trying again. This reinforces the engineering cycle. |
| **AFTER the Event** | Marshmallows will be all over the room, so make sure to check everywhere to get them all! |
| **Additional Resources** | Put out some meter sticks so students can measure how far they can launch their marshmallows. You might want to add materials, such as craft sticks, bottle caps, and pencils for students to experiment with other designs. |
| **Things to Think About** | 1. What causes your marshmallow to move? (The force applied to the spoon stores energy, which is transferred to the marshmallow when the spoon is released.)<br>2. Does the placement of the rubber band change the direction of the marshmallow?<br>3. Will adding or reducing the number of rubber bands change the force of your marshmallow?<br>4. Does how far you pull back on your catapult affect how far your marshmallow goes?<br>For all these questions, students should explore to determine variables. |

## Activity: CATAPULTS

### Program Information Section
*List location and hosts' names*

| | |
|---|---|
| **Science Behind the Activity** | Catapults are devices that launch objects. An arm connected to a "bucket" is attached to a base. In this design a spoon acts as the arm and the bucket. To launch your marshmallow, force is applied to the arm by pulling down on it, and potential energy is stored until it is released. When the arm is released, energy is transferred to a load in the spoon, which launches it at a target. |
| **Vocabulary** | • Force—the push or a pull on an object<br>• Potential Energy—the stored energy in an object<br>• Kinetic Energy—energy of motion<br>• Prototype—the first attempt at designing something |
| **Safety** | Make sure testers wear eye protection. Observers are to be at least 10 feet from the launch site. Never aim the catapult at a person or animal. Only launch soft or light objects in this activity. Remind students not to eat any food used in this activity. |

**What You Need:**
- ☐ Mini cups
- ☐ Rubber bands
- ☐ Mini spoons
- ☐ Mini marshmallows
- ☐ Wall target or buckets
- ☐ Safety goggles

**How-to:**

1. Choose from the following objects to design a catapult: cup, spoon, rubber bands, marshmallows

2. Use the engineering design process to design a catapult that can launch a mini marshmallow.

3. Test your prototype.

4. Redesign to improve your catapult's performance.

5. Continue to make changes until it can launch the farthest, fastest, and highest marshmallows.

**Things to Think About:**

1. What causes your marshmallow to move?

2. Does the placement of the rubber band change the direction of the marshmallow?

3. Will adding or reducing the number of rubber bands change the force of your marshmallow?

4. Does how far you pull back on your catapult affect how far your marshmallow goes?

**Instructions and General Information**

**Estimated Activity Time:** 10+ minutes

| Teacher Tips: CATCH THE WAVE | |
| --- | --- |
| **Core Content** | Earth Materials and Systems |
| **Crosscutting Concepts** | Energy and Matter |
| **Science and Engineering Practices** | Developing and Using Models |

| | |
| --- | --- |
| **Guiding Question** | How do waves carry energy? |
| **BEFORE the Event** | If you are using recycled water bottles, sterilize the mouth piece. Purchase mineral oil by the gallon. Its density is 0.8 g/cm³, so it will float on the water. Canola oil can also be used, but since it isn't clear, it isn't as attractive. Put down table coverings to make cleanup easier. Plastic fish toys should be more dense than mineral oil so that they are in or on the water. You may have to try several toy fishes to find one that will work. |
| **DURING the Event** | Students may need help sealing their bottles. The purpose of the balloon over the end is to prevent the lid from accidentally coming off or unscrewing. You might wish to use duct tape instead. |
| **AFTER the Event** | Make sure you clean and mop up all spills. |
| **Additional Resources** | You might want to add a station with a pan of water and straws for students to blow on to create waves. You might also want to add some glitter to make the water sparkle. Have some large wave bottles made from 2-liter bottles for students to explore while waiting. Consider using a nautical theme for decorating this room. |
| **Things to Think About** | 1. What happens to the water in the bottle when you tip it? (It creates a wave.) <br> 2. How is the energy you apply to your wave bottle different from the energy that makes waves in the ocean? (In the ocean, waves come from wind energy.) <br> 3. How can you create a bigger wave in your wave bottle? (Tip the bottle back and forth more, or faster.) |

In the DURING the Event diagram:

Mineral or baby oil

Make sure the cap is on tight!

Colored water

## Activity: CATCH THE WAVE

### Program Information Section
*List location and hosts' names*

| | |
|---|---|
| **Science Behind the Activity** | Water waves are formed by wind moving over water. Energy is transferred from the wind and carried by a wave through the water. Water waves are mechanical waves because they travel through a medium. The water doesn't actually travel with the wave, but moves up and down as the energy passes through. |

| | |
|---|---|
| **Vocabulary** | • Energy—power to make things move and change<br>• Mechanical waves—waves that travel through a medium<br>• Medium—material that waves travel through |
| **Safety** | Students should wear goggles in this activity. Sterilize the water bottles if using a recycled bottles. Have students wash hands after making their wave bottles. |

**What You Need:**

☐ Water bottle  ☐ Mineral oil
☐ Water  ☐ Funnels
☐ Food coloring  ☐ Balloons
☐ Toy fish or shells

**How-to:**

1. Fill or empty your bottle so that it is ¾ full of water.

2. Add blue food coloring.

3. Add a toy fish or shell.

4. Use a funnel to add mineral oil to fill the bottle.

5. Tighten the cap on the bottle.

6. Stretch a balloon over the cap to prevent it from coming loose or accidentally opening.

7. Add energy to your bottle by tipping it back and forth.

8. Watch the wave!

**Things to Think About:**

1. What happens to the water in the bottle when you tip it?

2. How is the energy you apply to your wave bottle different from the energy that makes waves in the ocean?

3. How can you create a bigger wave in your wave bottle?

**Instructions and General Information**

**Estimated Activity Time:** 10 minutes

# Chapter 10

| Teacher Tips: CHANGING-COLOR SLIME | |
|---|---|
| **Core Content** | Structure and Properties of Matter |
| **Crosscutting Concepts** | Stability and Change |
| **Science and Engineering Practices** | Constructing Explanations and Designing Solutions |

| | |
|---|---|
| **Guiding Questions** | What happens when different colors are combined? Can some substances be both a solid and a liquid? |
| **BEFORE the Event** | Cover tables where students will be working with slime. If you are working with small numbers, then students can combine the ingredients to make their own slime, allowing for a discussion of chemical changes. With large numbers, make the slime colors ahead of time. Use only primary colors (red, blue, and yellow). |
| **DURING the Event** | Students should choose two colors they want to mix. The food coloring can stain hands, so if you want to reduce mess, put small amounts of the slime in snack bags for mixing. |
| **AFTER the Event** | Make sure all tables, floors, and chairs are wiped off. |
| **Additional Resources** | For color mixing, you may want to allow students to paint with shaving cream that has been died with food coloring. Again, only use primary colors. If you want to demonstrate how light is different, have flashlights with blue, red and green cellophane over them in a dark corner or box for students to see how colors of light mix. |
| **Things to Think About** | 1. What color changes happen when you mix primary colors? Red and blue? Blue and yellow? Red and yellow? (Red and blue make purple, blue and yellow make green, and red and yellow make orange.) 2. What do you think would happen if you mixed primary and secondary colors? (When you mix primary and secondary colors, you get tertiary colors. If you have enough materials, you might allow students to explore this!) 3. Solids are materials that hold their shape. Liquids flow and take the shape of their container. Is your slime a solid or a liquid or something else? Explain! (Slime is a non-Newtonian fluid. It has properties of solids and liquids. Help students identify the characteristics of both solids and liquids for the slime.) |

## Activity: CHANGING-COLOR SLIME

### Program Information Section
*List location and hosts' names*

| | | |
|---|---|---|
| **Science Behind the Activity** | Primary colors are red, blue, and yellow. All colors can be made by mixing these three primary colors in different amounts. Secondary colors are made by mixing two different primary colors. This station uses slime, which is a non-Newtonian fluid, for mixing colors. Slime acts like a solid when it is pressed in your hands. When you open your fingers, it flows like a liquid. | **Color Wheel**<br><br>RED<br>purple   orange<br>BLUE   YELLOW<br>green |

| | |
|---|---|
| **Vocabulary** | • Primary colors—colors can be mixed to create all other colors. However, you cannot create primary colors mixing other ones.<br>• Secondary colors—colors that are produced by mixing two primary colors. They are green, orange, and purple.<br>• Non-Newtonian mixture—materials that don't behave like just one state of matter but have properties of solids and liquids. |
| **Safety** | Make sure table is clean. Advise children not to eat the slime. Students should wear safety goggles. |

### What You Need:

☐ Food coloring
☐ 2 cups of white glue
☐ 2½ teaspoons of borax
☐ Measuring cups and spoons

☐ 1½ cups of warm water plus 1 cup of warm water
☐ Plastic freezer bags—quart size and sandwich size
☐ Safety goggles

### How-to:

1. Mix in a quart-size freezer bag: 1½ cups warm water, 2 cups glue, and several drops of one color of food coloring—use primary colors only.

2. Mix in a sandwich bag or cup: 2½ teaspoons borax and 1 cup warm water.

3. Once both contents are completely mixed, pour the borax mix into the zip-lock with the glue mixture. Close bag and mix continually until slime is made.

4. Repeat using the other two primary colors.

### Things to Think About:

1. What color changes happen when you mix primary colors? Red and blue? Blue and yellow? Red and yellow?

2. What do you think would happen if you mixed primary and secondary colors?

3. Solids are materials that hold their shape. Liquids flow and take the shape of their container. Is your slime a solid or a liquid or something else? Explain!

**Instructions and General Information**

### Estimated Activity Time: 5+ minutes

# Chapter 10

| Teacher Tips: FOIL BOATS | |
|---|---|
| **Core Content** | Conservation of Energy and Energy Transfer |
| **Crosscutting Concepts** | Structure and Function |
| **Science and Engineering Practices** | Constructing Explanations and Designing Solutions |

| | |
|---|---|
| **Guiding Question** | How can you design a boat to hold the most weight possible? |
| **BEFORE the Event** | Place this activity in a room without carpet and where you have access to water (sink preferred). Make sure the water tubs are deep enough for the boats to sink but not too deep for smaller students to reach into. To help keep the water to a minimum on the tables, put the boats on a tray so the water can be emptied back into the buckets. Precut standard-sized foil sheets (or purchase foil already cut into sheets) prior to the event. |
| **DURING the Event** | The most important concept to emphasize is the engineering cycle. Have students make a prototype, then encourage them to re-engineer their boat to improve it. Make sure weights (or pennies) are dried between uses, as water has mass and wet weights will weigh more. You may want to use a whiteboard for tracking student outcomes in a "tournament" style, providing a prize to students who design the "best" boat at the end of the evening. |
| **AFTER the Event** | Make sure all the tables and floors are dry. Make sure weights are also dry when they are put away. |
| **Additional Resources** | Provide some plastic toy boats for students to see different shapes and styles of floating boats. Use a nautical theme to decorate this room! |
| **Things to Think About** | 1. How much weight did your boat hold? (Answers will vary. Encourage children to redesign.) <br> 2. Were you able to improve it so that it held more weights? How? (Answers will vary.) <br> 3. Does the height of the sides make a difference in how much weight a boat can hold? (Based on the design, students will have different answers.) <br> 4. Why do the weights float when placed in your boat but sink when put directly in the water? (Because the boat displaces more water than the weights do.) |

## Activity: FOIL BOATS

### Program Information Section
*List location and hosts' names*

| | |
|---|---|
| **Science Behind the Activity** | Buoyancy is how well something floats. There are two forces at work in determining if something floats. The first is gravity, which pulls things down. In this activity, gravity is trying to pull the foil and weights downward. The other force is buoyancy, which pushes your boat upward. When the force of buoyancy is greater than the force of gravity, your boat floats! For your boat to be buoyant, it must "displace" enough water so that the weight of the water is greater than the weight of the boat and its cargo. |
| **Vocabulary** | • Force—the push or pull on an object<br>• Gravity—the downward force on an object<br>• Buoyancy—the force that pushes objects up in water and causes them to float<br>• Displace—to move something |
| **Safety** | The floors can become slippery, so make sure you have mops and other materials to keep floors dry. Immediately wipe up any spilled water. Also keep a safe distance from electrical outlets and wires to protect again shock hazards. |

**What You Need:**

☐ Foil (cut into sheets)          ☐ Gram weights or pennies

☐ Tubs for water                     ☐ Rulers

**How-to:**

1. Use your imagination and problem-solving skills to design a boat that will float and hold the most weight possible.

2. Test your boat by placing it in the water and adding weights, one at a time, until it sinks.

3. Redesign to improve your boat and test it again.

**Things to Think About:**

1. How much weight did your boat hold?

2. Were you able to improve it so that it held more weights? How?

3. Does the height of the sides make a difference in how much weight a boat can hold?

4. Why do the weights float when placed in your boat but sink when put directly in the water?

**Instructions and General Information**

**Estimated Activity Time:** 10 minutes

| Teacher Tips: GETTING BUGGY | |
|---|---|
| Core Content | Natural Selection and Adaptation |
| Crosscutting Concepts | Structure and Function |
| Science and Engineering Practices | Constructing Explanations and Designing Solutions |

| | |
|---|---|
| **Guiding Question** | How does camouflage help animals survive? |
| **BEFORE the Event** | We found that purchasing "fat quarters" of fabric (quilting squares) was easier than cutting fabric and gave us more variety. Be sure to cut copy paper in half (hamburger style) for students. Purchase plastic insects in bulk. |
| **DURING the Event** | Suggest students try different combinations of colors and shapes. |
| **AFTER the Event** | Make sure none of the bugs have escaped! |
| **Additional Resources** | Decorate the room in a camouflage theme. Have photos of insects in a variety of habitats to show both camouflage and mimicry. |
| **Things to Think About** | 1. In which insect and fabric combination did the bug stand out the most? (The combination with the most contrast should make the bug stand out the most.)<br>2. In which insect and fabric combination did the bug stand out the least? (The combination with the least contrast should cause the bug to stand out the least. Also point that out the patterns and variety of colors on an insect can help it blend into a larger variety of backgrounds.)<br>3. Why did some insects stand out more than others? (Because they don't match their background.)<br>4. Do you think shape or color is more important in camouflage? Why? (Answers will vary. Encourage students to explain their responses and cite evidence from the activity.) |

## Activity: GETTING BUGGY

### Program Information Section
*List location and hosts' names*

| | |
|---|---|
| **Science Behind the Activity** | In the natural world, animals have many different adaptations that help them survive. In cold regions, some animals will hibernate during the winter to conserve energy. Because of drought and other seasonal changes, some animals migrate from one place to another to get the resources they need to survive. Some animals can hide from predators really well by blending in with their surroundings. Camouflage also improves an insect's ability to catch its prey! |
| **Vocabulary** | • Camouflage—an adaptation that allows an animal to blend in with its background<br>• Adaptations—physical characteristics that help animals survive<br>• Predators—organisms that live by hunting and eating others<br>• Prey—organisms that are hunted by predators |
| **Safety** | Make sure children don't try to put the bugs in their mouth. |

**What You Need:**

☐ Plastic insects

☐ Scraps of fabric

☐ Crayons or colored pencils

☐ Sheets of plain paper

**How-to:**

1. Place different insects on different pieces of fabric.

2. Try placing the same insect on different pieces of fabric.

3. Experiment with different combinations.

4. Select one bug to take home. On a piece of paper, draw and color an environment for your insect in which it will be safest.

**Things to Think About:**

1. In which insect and fabric combination did the bug stand out the most?

2. In which insect and fabric combination did the bug stand out the least?

3. Why did some insects stand out more than others?

4. Do you think shape or color is more important in camouflage? Why?

**Instructions and General Information**

**Estimated Activity Time:** 10 minutes

| Teacher Tips: GRASSY PETS | |
|---|---|
| **Core Content** | Cycles of Matter and Energy Transfer in Ecosystems |
| **Crosscutting Concepts** | Structure and Function |
| **Science and Engineering Practices** | Planning and Carrying Out Investigations |

| | |
|---|---|
| **Guiding Questions** | What are the needs of a plant? What are the parts of a plant? |
| **BEFORE the Event** | Purchase potting soil and grass seeds. We recommend throwing in a few radish seeds to make sure that students will grow all the parts of a plant, including leaves and flowers. Use clear cups so they will be able to observe the roots as they grow. Plant several grassy heads ahead of time to display. Also, provide a chart with the record of the height as an example of how to track the growth of the plant. |
| **DURING the Event** | Have students decorate a face on the cup. Poke holes in the bottom of the cup with a pushpin.<br><br>Do not fill the cup more than ¾ full of soil. Make sure to use a generous amount of grass seed to get a nice full "head" of hair. |
| **AFTER the Event** | Make sure all soil and seeds are cleaned up from tables and the floor. |
| **Additional Resources** | Display posters that show plants, plant needs and parts of a plant. You might wish to use a meadow theme to decorate this room. |
| **Things to Think About** | 1. What do plants need to grow? (Soil, water, light, air, space to grow.)<br>2. What part of the plant do you think will grow first? (The roots come out of the seed first.)<br>3. What part of the plant do you think will grow next? (The grass will grow next.)<br>4. What are the parts of a plant. (Root, stem, leaf, flower.) |

| Activity: GRASSY PETS |
|---|

**Program Information Section**
*List location and hosts' names*

| Science Behind the Activity | Like people, plants need certain things to survive. They need soil, water, air, nutrients, space, light, and warmth. The better these needs are met, the better plants will grow. Growing plants have different parts. The roots grow underground. The stem is the upper part of the plant that holds the leaves and flowers. In this activity, you will plant some seeds and take them home to see how they grow. | **Parts of a Plant**<br><br>Flower<br>Stem<br>Leaf<br>Roots |
|---|---|---|
| **Vocabulary** | • Nutrient—something that is needed for healthy growth | |
| **Safety** | If soap and water are not available, have antibacterial hand wipes for students to use after planting their seeds. Safety goggles are a must for this activity. Caution students not to put seeds in their mouths. Use caution when working with scissors and pushpins, which can cut or puncture skin. | |

**What You Need:**

- ☐ Potting soil
- ☐ Small plastic cups with lids
- ☐ Grass seeds
- ☐ Radish seeds
- ☐ Markers, colored paper, scissors
- ☐ Safety goggles
- ☐ Not necessary but fun: wiggly eyes and glue
- ☐ Pushpins

**How-to:**

1. Decorate your cup. Give it a face including eyes, a nose, and a mouth.

2. Poke small holes in the bottom of the cup with a pushpin.

3. Fill cup ¾ full with potting soil.

4. Sprinkle grass seeds on top. Add a few radish seeds.

5. Put lid on cup.

6. When you get home, remove lid and put under the cup, then place in a sunny window. Add a small amount of water daily to keep soil moist.

7. When you get home, design an experiment to try to get your grassy pet to grow the most. Keep a journal showing your plant's growth.

**Things to Think About:**

1. What do plants need to grow?

2. What part of the plant do you think will grow first?

3. What part of the plant do you think will grow next?

4. What are the parts of a plant?

**Instructions and General Information**

**Estimated Activity Time:** 5+ minutes

# Chapter 10

| Teacher Tips: HARMONY HARMONICAS | |
|---|---|
| **Core Content** | Waves: Light and Sound |
| **Crosscutting Concepts** | Cause and Effect |
| **Science and Engineering Practices** | Planning and Carrying Out Investigations |

| | |
|---|---|
| **Guiding Question** | How is sound produced? |
| **BEFORE the Event** | To make the event go easier, precut straws into 1 in. pieces. Put down table coverings to prevent markers from staining desks. |
| **DURING the Event** | Small children may lack the fine motor skills needed to assemble their harmonicas. They may need some help with the rubber bands and straws. |
| **AFTER the Event** | Make sure to sweep up all small pieces of straws and rubber bands. |
| **Additional Resources** | Have a variety of musical instruments that students can explore. Take slow-motion videos of their harmonicas to see the vibrations. |
| **Things to Think About** | 1. What causes the instrument to make sounds? (Vibrations of the rubber band against the stick.)<br>2. What happens when you move the position of the straws? (The pitch changes.) |

National Science Teachers Association

## Activity: HARMONY HARMONICAS

### Program Information Section
*List location and hosts' names*

| | |
|---|---|
| **Science Behind the Activity** | Sound is created when something vibrates. Blowing on your harmonica makes the rubber band vibrate against the stick, which then moves air molecules to create sound. |
| **Vocabulary** | • Vibration—rapid back-and-forth motion<br>• Pitch—how high or low a sound is<br>• Harmony—pleasant combination of sounds |
| **Safety** | Caution students not to snap rubber bands at each other. Watch to see that small students don't put pieces in their mouths. Use caution when using scissors, which can cut or puncture skin. |

**What You Need:**

☐ Craft sticks          ☐ Scissors
☐ Rubber bands       ☐ Markers or colored pencils
☐ Straw

**How-to:**

1. Cut two straw pieces, about 2 in. long.

2. Place one rubber band around one craft stick (the long way).

3. Put the two straw pieces under the rubber band, one at each end.

4. Place the second stick over the first stick and straw assembly.

5. Wrap the ends with the remaining rubber bands.

6. Decorate your harmonica!

7. Blow! See if you can find a way to get your harmonica to make different pitches.

**Things to Think About:**

1. What causes the instrument to make sounds?

2. What happens when you move the position of the straws?

**Instructions and General Information**

**Estimated Activity Time:** 10+ minutes

| Teacher Tips: ICE CREAM | |
|---|---|
| **Core Content** | Conservation of Energy and Energy Transfer |
| **Crosscutting Concepts** | Energy and Matter |
| **Science and Engineering Practices** | Planning and Carrying Out Investigations |

| | |
|---|---|
| **Guiding Questions** | How does ice cream freeze? <br> Why does it melt? |
| **BEFORE the Event** | This is a fun but messy event! We recommend driveway salt for winter weather, as it is less expensive than rock salt sold for ice cream in the grocery store. There is always a crowd at this activity so have a space with lots of room. A good location is the lunch room to make it easier to manage. You will need an ice machine nearby. Put down tablecloths for easier cleanup. Have a mop and bucket to clean up spills. Designate one area (over a table is best) to shake the bags. Have garbage pails for disposing of trash and sinks or bins for melted ice. |
| **DURING the Event** | We suggest having students use straws instead of spoons to eat their ice cream. It keeps the spills to a minimum. Encourage students to take the inside bag out to eat their ice cream if possible. Dump the salt-ice mix from the big bags into buckets, and throw the large freezer bags in the garbage. If you don't dump the ice out first, the garbage bags will be too heavy to manage. |
| **AFTER the Event** | Make sure all tables are cleaned off. Warm, soapy water is best to get all the salt off the tables. You'll probably need to mop the floor more than once, or ask the custodian to use the floor machine to make sure all sticky residue is cleaned up. |
| **Additional Resources** | Posters explaining the states of matter. |
| **Things to Think About** | 1. What part of the activity is a solid? (Ice cream, straw, bag, ice, sugar, salt.) A liquid? (Milk, vanilla.) A gas? (Air in and around the bag.) <br> 2. What do we call the phase change from a liquid to a solid? (This phase change is called freezing.) <br> 3. How can the solid go back to being a liquid? (When the heat from the air and your hands causes the ice cream to melt.) |

## Activity: ICE CREAM

### Program Information Section
*List location and hosts' names*

| | |
|---|---|
| **Science Behind the Activity** | There are three basic states of matter—solids, liquids, and gases. When heat is added or taken away, objects transition from one state to another. These phase changes are the melting or freezing point and the boiling or condensation point of a substance. When salt is added to water, it lowers the melting point, and salt water can actually get colder than freezing! This allows the milk mixture in the small plastic bag to freeze into solid ice cream. While making your ice cream, you can observe a phase change—a liquid freezing and changing into a solid! |

(diagram: Temperature vs. Energy Added, showing SOLID, Phase Change: Melting or Freezing, LIQUID, Phase Change: Boiling or Condensing, GAS)

| | |
|---|---|
| **Vocabulary** | • Solid—the state of matter where atoms are tightly packed together so that they have a defined shape<br><br>• Liquid—the state of matter where there is enough space between atoms that they will move and take the shape of their container<br><br>• Gas—the state of matter where there is enough space between atoms that they can spread out to fill up a space<br><br>• Phase change—when enough energy is added to or taken away from an object that it moves from one state (solid, liquid, or gas) to another |
| **Safety** | Make sure to keep floors dry to prevent slipping. |

**What You Need:**

- ☐ 1 cup milk
- ☐ ½ teaspoon vanilla
- ☐ 2 tablespoons sugar
- ☐ ½ cup rock salt
- ☐ 2 cups ice
- ☐ 1 quart-size plastic freezer bag
- ☐ 1 gallon-size plastic freezer bag
- ☐ 1 straw

**How-to:**

1. Put the milk, sugar, and vanilla into the quart-sized bag. Seal the bag.
2. Put the bag into the gallon-sized bag.
3. Add ice and rock salt. Seal the bag.
4. Shake the bag gently until it forms ice cream!

**Things to Think About:**

1. What part of the activity is a solid? A liquid? A gas?
2. What do we call the phase change from a liquid to a solid?
3. How can the solid go back to being a liquid?

**Instructions and General Information**

**Estimated Activity Time:** 10+ minutes

| Teacher Tips: MARACAS | |
|---|---|
| Core Content | Conservation of Energy and Energy Transfer |
| Crosscutting Concepts | Cause and Effect |
| Science and Engineering Practices | Planning and Carrying Out Investigations |

| | |
|---|---|
| Guiding Questions | How is sound produced? <br><br> Can you make an instrument that produces sound? |
| BEFORE the Event | Store the beans, popcorn, and rice in separate containers with lids to keep them dry. Use fun duct tape to make the maracas festive. Make sample maracas for the students to try to determine what items they want inside the maracas they will make. |
| DURING the Event | To move more people through quickly, have two separate stations set up with the same items. |
| AFTER the Event | Sweep the floor to make sure all the beans and rice are off the floor. |
| Additional Resources | Provide real maracas of different shapes and sizes for students to play with while they are waiting their turn. Have other percussion instruments available for students to play with. |
| Things to Think About | 1. What do you hear? (When you shake the egg, the entire instrument is vibrating, thus creating the sound.) <br><br> 2. How is the sound produced? (When the materials inside the maraca hit each other, they create vibrations. The plastic egg transfers some of that sound energy to the surrounding air, which travels to your ear.) <br><br> 3. How can you change the sound? (Change the items that are inside the egg. The material, weight, and number of objects inside can change the pitch of the sound you hear.) |

## Activity: MARACAS

### Program Information Section
*List location and hosts' names*

| | |
|---|---|
| **Science Behind the Activity** | Sound is produced by vibration. Maracas are a kind of percussion instrument called an idiophone, which makes sound by vibrating the entire instrument. Those vibrations are transferred to air and produce the sound we hear. Different objects put inside a maraca will produce sounds with different pitches. |

**Fill egg** →

**Duct tape egg between spoons**

| | |
|---|---|
| **Vocabulary** | • Sound—a form of energy that travels in invisible waves<br>• Medium—the material the sound waves travel through (solid, liquid, gas)<br>• Percussion—a type of instrument that produces sounds by being struck |
| **Safety** | Monitor young children with the small objects that are used to make the maracas (e.g., beans, rice). Make sure they don't eat them or put them in their mouths, ears, or noses. |

**What You Need:**

☐ Plastic eggs
☐ Rice or beans
☐ Plastic spoons
☐ Duct tape

**How-to:**

1. Fill one of the halves of the plastic egg with beans or rice.
2. Close the egg.
3. Use duct tape to secure the egg between two spoons.

**Things to Think About:**

1. What do you hear?
2. How is the sound produced?
3. How can you change the sound?

**Instructions and General Information**

**Estimated Activity Time:** 5+ minutes

| Teacher Tips: PINWHEELS | |
|---|---|
| Core Content | Earth's Systems |
| Crosscutting Concepts | Scale, Proportion, and Quantity |
| Science and Engineering Practices | Developing and Using Models |

| | |
|---|---|
| **Guiding Question** | What causes the pinwheel to spin? |
| **BEFORE the Event** | Pre-punch the holes in the straws. Copy the pinwheel template on card stock or heavy paper. We've also used transparency film in the past to produce more durable pinwheels. Provide crayons and colored pencils for the students to decorate their pinwheels while waiting their turn to put them together. |
| **DURING the Event** | It is important to keep the brad loose when putting everything together so the wheel will move freely. If it does not move well, ask the students to problem solve why before fixing it. |
| **AFTER the Event** | Sweep up all the punches from the paper and the straws. |
| **Additional Resources** | Have pictures of weather instruments that measure wind speeds. Also provide pictures of wind farms. Another variation is to use pushpins in the tip of an eraser instead of brads and beads. |
| **Things to Think About** | 1. What can you do to increase the number of turns (revolutions) of the pinwheel? (Blow it harder to increase the "wind" speed.) 2. How can a pinwheel be used in weather observations? (By observing the number of times the sticker goes around, you can see if the wind is increasing or decreasing.) 3. In what other ways can this type of machine help us? (Windmills on wind farms are called turbines. These machines use wind energy to turn a generator to produce electricity.) |

## Activity: PINWHEELS

### Program Information Section
*List location and hosts' names*

| | |
|---|---|
| **Science Behind the Activity** | Pinwheels are designed to capture moving air. When the wind blows into the pinwheel, the blades spin. This design is used in windmills and turbines to convert wind energy into electrical or mechanical energy. Many different machines use spinning blades. |

| | |
|---|---|
| **Vocabulary** | • Windmill—converts wind energy to mechanical energy<br>• Wind turbine—converts wind energy to electrical energy (electricity) |
| **Safety** | Be sure to keep paper fasteners away from small children, as they are choking hazards. |

**What You Need:**

☐ Pinwheel template
☐ Cardstock or heavy paper
☐ Scissors
☐ 2 in. paper fasteners (brads)
☐ Beads
☐ Colored pencils
☐ Extra-wide straws (smoothie straws work well) with a hole punched in the top
☐ Dot stickers in a dark color

**How-to:**

1. Cut out the pattern and cut on the lines.
2. Decorate using colored pencils.
3. Punch holes in each of the corners and in the center where indicated.
4. Fold the corner holes to the middle and line them up.
5. Put a bead on the brad, then push the brad through the lined-up holes.
6. Put on another bead, and push the brad through the hole in the straw.
7. Open the brad in the back to keep the bead on but loose enough that the pinwheel can turn freely.
8. Put a colored dot on one of the folds (dark color sticker works well).

**Things to Think About:**

1. What can you do to increase the number of turns (revolutions) of the pinwheel?
2. How can a pinwheel be used in weather observations?
3. In what other ways can this type of machine help us?

**Instructions and General Information**

**Estimated Activity Time:** 10 minutes

# Chapter 11

# Intermediate-Level Activities

Here you will find activities that we believe to be at an intermediate level. Either the concepts are more complex than those in the previous chapter or they require more complex motor skills. Choose several activities from this chapter for your event, making sure that you pull from a variety of concepts. For activities with similar concepts, trade off these activities by year so that attendees get a different experience. Also remember that by asking different questions, you can use the same activities at different levels. So if you have older students at these activities, ask more difficult questions. The activities presented here are as follows:

- Air Cannon—build an air cannon and use air to move a target
- "Bear-y" Hot S'mores—explore thermal energy by making this favorite treat
- Bug Buzzers—create sound with vibrations from insects made out of cardstock
- Dino Discovery—dig for dinosaurs and create fossils with plaster
- Face Magnets—explore magnetism by making a toy
- Lava Lamps—learn about the different densities of liquids
- Parachutes—participate in an engineering challenge to make the best parachute
- Plant Dissection—discover the parts of a flower and seed as they are dissected
- Roller Coaster Challenge—engineer a roller coaster with loops and turns
- Static Sensations—discover the world of static electricity
- Straw Rockets—explore forces and motion by building a straw rocket
- Surface Tension—learn about cohesion of water and make a water strider

| Teacher Tips: AIR CANNON | |
|---|---|
| **Core Content** | Conservation of Energy and Energy Transfer |
| **Crosscutting Concepts** | Energy and Matter |
| **Science and Engineering Practices** | Planning and Carrying Out Investigations |

| | |
|---|---|
| **Guiding Question** | How can air be used to move something? |
| **BEFORE the Event** | Remove the bottom ends of your plastic cups. Cut balloon off at the bottom of the neck. You will want to use latex-free balloons to avoid allergic reactions. Cut 3 in. discs out of file folders with a 1 in. hole in the middle of each disc. You can get hole punchers these sizes from an office supply store to make cutting fast and easy. Have these precut to make assembly easy during the event. Assemble samples of the air cannon ahead of time for students to see the end product. Set up an area on the table with paper confetti. Have extra confetti paper to replace it during the night. Bring broom and dustpan to clean up paper on the floor. |
| **DURING the Event** | When many students are coming through, it is faster to have the tape precut and hanging on the side of the table. You may have to help young children stretch the balloon over the wide end of the cup. Students will practice using their air cannons on the confetti. |
| **AFTER the Event** | Make sure to clean up all the paper confetti from the floor. |
| **Additional Resources** | Large air cannons can be made or purchased for students to experiment with while waiting to create ones of their own. |
| **Things to Think About** | 1. Identify the pushing and pulling forces when you use your air cannon. (Push—the air being forced out of the cup. Pull—when you are pulling back on the balloon.)<br>2. Does the amount of force you use to pull back the balloon affect how the confetti is pushed when it the balloon is released? (The more air you can put in your air cannon when you pull back, the greater the force on the confetti.) |

## Activity: AIR CANNON

### Program Information Section
*List location and hosts' names*

| | |
|---|---|
| **Science Behind the Activity** | Air has mass and takes up space. When the balloon of an air cannon is pulled back, the volume of air inside increases. When the balloon is released, the amount of air decreases rapidly. In the center of the cannon, air moves out faster, and on the edges, it moves more slowly, creating an air "doughnut" or vortex. When you launch an air cannon, the air that is released can travel over a surprisingly long distance! |

*Balloon stretched over large end of cup*

*Disc with hole in center taped to small end of cup*

| | |
|---|---|
| **Vocabulary** | • Force—push or a pull on an object<br>• Push—to apply force to move something away<br>• Pull—to apply force to bring something closer |
| **Safety** | To avoid allergic reactions to latex, use latex-free balloons. Always have students wear protective eyewear when using projectiles. |

**What You Need:**

☐ Latex-free balloons—cut below the neck
☐ Plastic cups—with the bottom cut off
☐ Tape—any type
☐ Scissors
☐ 3 in. circles cut from file folders, with 1 in. circle cut in center of each
☐ Paper confetti
☐ Safety goggles

**How-to:**

1. Attach the disc to the small (bottom) end of the cup with tape.

2. Stretch the balloon over the large end of the cup.

3. Point the air cannon at some confetti on the table, pull back on the balloon, and release.

**Things to Think About:**

1. Identify the pushing and pulling forces when you use your air cannon.

2. Does the amount of force you use to pull back the balloon affect how the confetti is pushed when the balloon is released?

**Instructions and General Information**

**Estimated Activity Time:** 10 minutes

# Chapter 11

| Teacher Tips: "BEAR-Y" HOT S'MORES | |
|---|---|
| **Core Content** | Conservation of Energy and Energy Transfer |
| **Crosscutting Concepts** | Energy and Matter |
| **Science and Engineering Practices** | Analyzing and Interpreting Data |

| | |
|---|---|
| **Guiding Question** | How does energy cook food? |
| **BEFORE the Event** | Heat lamps work well for this activity and may be borrowed from a local restaurant. Be sure your hosts know how to operate them. Determine how far the lamp needs to be from the s'more in order for the chocolate to melt. Place tape on the floor to create a line to keep small children away from the heat lamps while their s'mores cook. Be sure to put down a protective layer of cardboard over the table so that the lamps do not damage it. You can use traditional alcohol thermometers, but we recommend using a thermal gun so that students can collect temperature data continuously while their s'mores cook. |
| **DURING the Event** | Collect temperature data by putting one thermometer inside the tin foil with the s'more and one on the table, outside the heated area. Or if using a thermal gun, allow students to take the temperature of the s'more as it cooks. The lamps should be operated by the hosts. Take care to keep small children back so that they do not get hurt. Make sure they can still see their 's'mores to observe the melting. |
| **AFTER the Event** | Unplug lamps while cleaning up so they have time to cool before they are packed. Make sure all food is cleaned up from tables and floor. |
| **Additional Resources** | Have extra thermal guns for students to collect data of different objects in the room. You might also want to provide a child's bath thermometer and tubs of water of different temperatures for students to explore thermal energy. |
| **Things to Think About** | 1. What happened to the temperature of the s'more while it cooked? (The temperature should have gone up.)<br>2. How did the temperature of the s'more compare with the temperature of the thermometer that was away from the lamp? (It should have gotten hotter under the lamp.)<br>3. How does the heat get from the lamp to the s'more? (The heat radiated from the lamp.)<br>4. What makes the chocolate on your s'more melt? (Heat from the lamp.) |

## Activity: "BEAR-Y" HOT S'MORES

### Program Information Section
*List location and hosts' names*

| | |
|---|---|
| **Science Behind the Activity** | Heat is how energy is transferred from one object to another because of a difference in temperature. A lamp provides thermal energy in the form of heat, which is transferred to the s'more. Heat is always transferred from hot to cold objects. When an object is heated, the molecules in it speed up. Heat can be transferred in three different ways: conduction, convection and radiation. The type of heat transferred by lamps is radiation, which means it is heat that can travel through air or space. |
| **Vocabulary** | • Heat energy—also called thermal energy, is the energy an object has because of the movement of its molecules<br>• Temperature—a measure of heat energy |
| **Safety** | Purchase some gluten-free Teddy Grahams or graham crackers for special dietary needs. Make sure lamps are placed where students cannot touch them. Make sure s'mores are not too hot so that children are not burned. |

**What You Need:**

☐ Teddy Grahams ☐ Napkins
☐ Chocolate chips ☐ Tin foil
☐ Mini marshmallows ☐ Thermometers or thermal gun
☐ Heat lamp ☐ Permanent marker

**How-to:**

1. Take two Teddy Grahams. Put a marshmallow and a chocolate chip between them.

2. Wrap this in tin foil, and use a permanent marker to write your name on it.

3. Place it under the heat lamp with the help of an adult.

4. Take the temperature of your s'more while it cooks under the lamp. Compare it with the temperature of the table away from the lamp.

5. Enjoy!

**Things to Think About:**

1. What happened to the temperature of the s'more while it cooked?

2. How did the temperature of the s'more compare with the temperature of the thermometer that was away from the lamp?

3. How does the heat get from the lamp to the s'more?

4. What makes the chocolate on your s'more melt?

**Instructions and General Information**

**Estimated Activity Time:** 10+ minutes

| Teacher Tips: BUG BUZZERS | |
|---|---|
| Core Content | Sound Waves |
| Crosscutting Concepts | Cause and Effect |
| Science and Engineering Practices | Planning and Carrying Out Investigations |

| | |
|---|---|
| **Guiding Questions** | How do vibrations cause sound?<br>How can you change the pitch of the sound? |
| **BEFORE the Event** | Prepare your materials by precutting the string. Tape the pieces of string on the sides of the table to keep them from getting tangled. We found that thinner string (kite string) works better than yarn. Set up an area for children to experiment with their bug buzzers. |
| **DURING the Event** | Make sure the string is taped onto the end of the craft stick before the eraser is put on. The string will come out from under the eraser. Put the rubber band on last, making sure the string is not over the rubber band. If the bug buzzer does not buzz, make sure the rubber band is not twisted and that the string is not over the rubber band. Limit the number of students testing at one time to reduce accidents. |
| **AFTER the Event** | Sometimes buzzers can fly off and end up scattered all around the room. Make sure all are accounted for and the room is cleaned up. |
| **Additional Resources** | Pictures of various bug wing shapes, instruments that show vibration, and some homemade instruments that use rubber bands to create sound could be set up in a station for free exploration. |
| **Things to Think About** | 1. What makes the sound? (The rubber band vibrates against the stick to create the sound.)<br><br>2. What happens to the sound when you twirl your bug faster? Slower? (The pitch is higher when you spin it faster, lower when it is slower.)<br><br>3. How will changing the size of the rubber band affect the sound? (A thinner rubber band usually will make a higher pitch, and a thicker rubber band will make a lower pitch.)<br><br>4. What other parts of your bug could be changed to affect the sound? (Try changing the size and shape of the wings.) |

## Activity: BUG BUZZERS

### Program Information Section
*List location and hosts' names*

| | |
|---|---|
| **Science Behind the Activity** | Sound is a wave produced by vibrating air. When you swing your bug buzzers, the rubber band vibrates against the stick, which causes the air around the rubber band to vibrate. Changing the speed at which you spin your bug buzzer produces a different pitch. |

Tape string under end before adding eraser

Rubber band OVER erasers and "bug"

Tape bug to craft stick

Erasers at ends

Tape at ends of card

String comes out at end from under eraser

| | |
|---|---|
| **Vocabulary** | • Sound—form of energy made by vibrations<br>• Pitch—how high or low a sound is |

| | |
|---|---|
| **Safety** | Goggles are a must to protect eyes from flying bug buzzers! Have students test their buzzers in an area away from other students. Make sure they have plenty of room so that they don't hurt others! |

**What You Need:**

☐ Pencil cap erasers    ☐ Index card    ☐ Scissors
☐ Tape    ☐ Craft stick    ☐ Colored pencils
☐ String    ☐ Rubber bands    ☐ Safety goggles

**How-to:**

1. Cut an index card to form a "bug" wing. You can decorate it if you want!
2. Tape the string to one end of the craft stick.
3. Tape the edges of the bug wing to the popsicle stick.
4. Put an eraser cap on the end of the stick.
5. Put a rubber band over the cap erasers.
6. Holding the string, swing the buzzer around in a circle and hear the bug buzzzzzz!

**Things to Think About:**

1. What makes the sound?
2. What happens to the sound when you twirl your bug faster? Slower?
3. How will changing the size of the rubber band affect the sound?
4. What other parts of your bug could be changed to affect the sound?

**Instructions and General Information**

**Estimated Activity Time:** 10+ minutes

| Teacher Tips: DINO DISCOVERY | |
|---|---|
| Core Content | Evidence of Common Ancestry and Diversity |
| Crosscutting Concepts | Stability and Change |
| Science and Engineering Practices | Engaging in Argument From Evidence |

| | |
|---|---|
| Guiding Question | How do fossils form? |
| BEFORE the Event | Set up a sandbox for the first station in this activity. Bury small, plastic dinosaurs before students arrive. Include a variety of sifting tools students can use for their dig. All ages love to do this, so it is a good area to keep participants busy when it gets crowded. Make sure there is a sink in this room, and preferably, this activity will be located in a room without carpet. Be sure to cover the tables where you will be working with plaster. |
| DURING the Event | Let students dig for dinosaur "fossils" at the first station. They will take their finds to the plaster station where they will create a cast "fossil" of the dinosaur in plaster. Sometimes the plaster can be hard to work with, especially if there is not be a good water source. A great alternative is to use clay or playdough instead of plaster. Make sure to have a broom and dustpan to keep the area clean during event. |
| AFTER the Event | Make sure all sand and plaster are cleaned up from tables, chairs, and floors. |
| Additional Resources | Have a variety of objects for students to choose to make fossils from. These could be shells, teeth, feathers, and so on. Put up a display of different kinds of fossils. Hang posters explaining the formation of fossils. Arrange to set up some real fossils at a station for students to explore. |
| Things to Think About | 1. What type of fossil did you make in the plaster? (A mold fossil.) 2. What type of fossil was made with the clay? (A cast fossil.) 3. How could you make a trace fossil? (One way would be to make footprints of the dinosaur in the plaster.) 4. What can you tell about things that lived long ago from fossils? (Fossils tell us about the things that lived in the past.) |

## Activity: DINO DISCOVERY

### Program Information Section
*List location and hosts' names*

| | |
|---|---|
| **Science Behind the Activity** | Fossils are remains and traces of plants and animals that lived long ago. When a plant or animal dies, it can leave an impression in sediment before it decays that forms a mold. Later, when the mold fills with minerals, a cast is created. Both the cast and the mold are types of fossils. Sometimes, fossils are formed from impressions left in sediment that come from animal activities, rather than their tissues. These are trace fossils. Fossils tell a story about plants and animals that lived when the fossil formed. |
| **Vocabulary** | • Cast—a type of fossil that is formed when a liquid substance (minerals, over time) is poured into a form or mold to form an impression<br>• Mold—a type of fossil formed when an organism dies and decays but leaves a hollow impression in the ground<br>• Trace—a type of fossil that is an impression from activities of ancient animals (e.g., footprints, scratches, burrows) |
| **Safety** | Students should wear goggles when working with plaster. The floor can become slippery when the sand is spilled. Students should wash hands with soap and water upon completing this activity. |

**What You Need:**

☐ Plastic or paper cups  ☐ Small, plastic dinosaurs  ☐ Water
☐ Plastic spoons  ☐ Sifting tools  ☐ Clay
☐ Plaster  ☐ Large container for  ☐ Safety goggles
☐ Sand     "dig"

**How-to:**

1. Dig in the sand to find a dinosaur fossil.
2. Take your dinosaur fossil to the plaster station.
3. Fill ¼ of a cup with plaster and add water (two parts plaster to one part water).
4. Stir, then wait about two to three minutes.
5. Make an impression in the plaster with the dinosaur.
6. After the plaster dries, put some clay into the impression left by the dinosaur.

**Things to Think About:**

1. What type of fossil did you make in the plaster?
2. What type of fossil was made with the clay?
3. How could you make a trace fossil?
4. What can you tell about things that lived long ago from fossils?

**Instructions and General Information**

**Estimated Activity Time:** 10 minutes

| Teacher Tips: FACE MAGNETS | |
|---|---|
| **Core Content** | Forces and Interactions |
| **Crosscutting Concepts** | Cause and Effect |
| **Science and Engineering Practices** | Planning and Carrying Out Investigations |

| | |
|---|---|
| **Guiding Questions** | How do magnets interact with other objects?<br>How can magnetism make iron filings move? |
| **BEFORE the Event** | Prepare face cards in advance by laminating cardstock. You can have premade faces, or have plain sheets for students to draw their own faces. Make sure that when you cut the face cards, they will fit into the plastic sleeves. Use table covers to reduce the mess on desks. Put the iron filings in a bowl to make it easier for the students to scoop them out. |
| **DURING the Event** | Spoons can be used to measure the correct amount of iron filings. You may want to use small funnels to make sure the filings get into the holders without spilling. Children may need help sealing the sleeves. We recommend putting a piece of clear packing tape over the opening end to prevent students from losing their iron filings. |
| **AFTER the Event** | Make sure to sweep up any iron filings from tables and floors. |
| **Additional Resources** | Set up a center with purchased magnetic face toys for students to play while they are waiting. Have a second center with wand magnets and a variety of objects for students to explore magnetism. |
| **Things to Think About** | 1. Why do the iron filings move around the face? (The magnetic force attracts them to the magnet on the wand.)<br>2. Why do you think iron filings are used for this activity instead of something else, like sand? (Because the iron filings are magnetic.)<br>3. In this activity, is magnetism a pushing or pulling force? (It is a pulling force.)<br>4. What happens when you take the wand away from the iron filings? (They stop moving. If the face is held upright, the force of gravity will pull them back to the bottom of the sleeve [they will fall].) |

## Activity: FACE MAGNETS

### Program Information Section
*List location and hosts' names*

| | |
|---|---|
| **Science Behind the Activity** | Magnetism is a force of attraction that acts on certain metals. When the magnet on the spoon comes close to the iron filings, the filings are attracted to the magnet and will move with it. When the magnet is pulled away, the filings lose the attraction and fall back down. |
| **Vocabulary** | • Force—a push or a pull on an object<br>• Repel—to push away from an object<br>• Attract—to pull or draw toward an object |
| **Safety** | Have students working with the iron filings wear goggles. Remind them never to use the magnets around electronics. |

**What You Need:**

☐ Iron filings     ☐ Clear packing tape     ☐ Clear name tag holders with zip-close top
☐ Plastic spoons     ☐ Permanent markers     ☐ Small round magnet with self-stick side
☐ Safety goggles     ☐ Laminated cardstock

**How-to:**

1. Draw a face on laminated cardstock with a permanent marker.

2. Place a card in the clear name tag holder.

3. Add iron filings.

4. Zip the top closed, then tape it to make sure it stays closed.

5. Put a self-sticking magnet on a spoon to make a magnetic wand.

6. Use the magnet to decorate the face with magnetic hairs!

**Things to Think About:**

1. Why do the iron filings move around the face?

2. Why do you think iron filings are used for this activity instead of something else, like sand?

3. In this activity, is magnetism a pushing or pulling force?

4. What happens when you take the wand away from the iron filings?

**Instructions and General Information**

**Estimated Activity Time:** 10 minutes

| Teacher Tips: LAVA LAMPS | |
|---|---|
| **Core Content** | Structure and Properties of Matter |
| **Crosscutting Concepts** | Cause and Effect |
| **Science and Engineering Practices** | Developing and Using Models |

| | |
|---|---|
| **Guiding Question** | How can density be used to separate fluids?<br>What happens when fluids of different densities are mixed together? |
| **BEFORE the Event** | Cover tables with tablecloths. It is best to assign this activity to a room with a sink. There are several container options for this activity. We recommend using baby soda bottles as test tubes with stands. Put the water and oil into condiment squeeze bottles or use funnels to make it easier to pour into tubes. Have buckets available for used liquids. Have a mop and bucket or paper towels ready to wipe up any spills. |
| **DURING the Event** | Help students pour the liquids if necessary to minimize spills. Students can choose any color of food coloring drops they want for their container.<br><br>You can reuse the tubes or containers if you want. Each group can pick a different color than is already in the water to see if they mix or stay separate.<br><br>Another fun way to observe the bubbles is to turn off the lights and hold a flashlight behind or under the tube! |
| **AFTER the Event** | Make sure to dispose of liquids from this activity according to school guidelines. |
| **Additional Resources** | You may wish to include some commercial wave machines that have fluids of different densities or "ooze tubes" to demonstrate. You may also choose to use a density column to demonstrate densities of fluids or density blocks to explore solids of different densities. |
| **Things to Think About** | 1. What direction are the bubbles traveling? Why? (They are going up because the carbon dioxide is lighter than both the liquids.)<br>2. Does that direction change? Why? (When the bubble pops, the water goes back down through the oil because it less dense than the oil.) |

## Activity: LAVA LAMPS

**Program Information Section**

*List location and hosts' names*

| | |
|---|---|
| **Science Behind the Activity** | Water and oil are liquids with different densities. Because oil is less dense than water, it will float on top of it. When you put both in the same test tube, they will form layers with the denser water sinking to the bottom. Food coloring is the same density as the water. When you put it on top of the oil, it will sink to the level of the water because it is the same density. When a fizz tablet is dropped in, it dissolves creating bubbles (carbon dioxide). This gas is even less dense than the liquids, so it goes to the top while trapping some of the colored water in the bubble. When the bubble pops at the top and releases the gas, the water goes back down.  |
| **Vocabulary** | • Density—how close together the molecules are packed |
| **Safety** | Have students wear safety goggles. Warn students not to drink liquids. Make sure to wipe up spills to prevent the floor from becoming slippery. |

**What You Need:**

☐ 8 oz. water bottles, soda bottle preforms (baby soda bottles), or large test tubes

☐ Vegetable oil

☐ Water

☐ Food coloring

☐ Fizz tablets (¼ tablet per container)

☐ Safety goggles

**How-to:**

1. Fill container ¾ of the way with oil and the rest of the way with of water.

2. Put about 3 drops of food coloring.

3. Drop in ¼ fizz tablet.

**Things to Think About:**

1. What direction are the bubbles traveling? Why?

2. Does that direction change? Why?

**Instructions and General Information**

**Estimated Activity Time:** 10 minutes

# Chapter 11

| Teacher Tips: PARACHUTES | |
|---|---|
| **Core Content** | Force and Motion |
| **Crosscutting Concepts** | Structure and Function |
| **Science and Engineering Practices** | Constructing Explanations and Designing Solutions |

| | |
|---|---|
| **Guiding Questions** | How does a parachute work?<br>What design will make the parachute slow down the most? |
| **BEFORE the Event** | You'll need a large area to test the parachutes. Set up a high place for testing, but only allow adult volunteers to test parachutes from this high point. You can use the second level in a gym or second story in an open floor plan in a school. If in a music room, use the risers to add height for testing. We had our activity hosts stand on chairs. |
| **DURING the Event** | The goal of this activity is to encourage creative thinking and engage students in the engineering cycle. Give students a variety of materials for both the parachute and weights so they can use the engineering cycle steps to design and then improve on their design. You can record the drop times of different students' parachutes and use the data at the end of the evening to declare a winner. |
| **AFTER the Event** | Make sure all the furniture is returned to original position. |
| **Additional Resources** | Provide pictures or videos of different types of parachutes. You might want to include some purchased parachute toys for students to play with. If possible, find someone who skydives to bring in his or her parachute with some pictures of him or her skydiving to talk to the students and their families. |
| **Things to Think About** | 1. What is making the parachute fall? (The force of gravity is pulling it toward the ground.)<br>2. What is slowing the parachute down? (The force of air resistance is slowing it down.)<br>3. How can you modify the parachute to make it fall more slowly? (The curved sides of the parachute help "catch" the air. The larger the parachute, the more air resistance it has, which will make it fall more slowly.)<br>4. How are parachutes helpful? (They slow down things that are falling.) |

## Activity: PARACHUTES

### Program Information Section
*List location and hosts' names*

| | |
|---|---|
| **Science Behind the Activity** | When objects fall to Earth, they have two forces acting on them: gravity and air resistance. Gravity is the force that pulls objects down, while air resistance works against gravity to slow down falling objects. Parachutes work because of the force of air resistance. As the parachute falls through the air, the air molecules under the parachute create air resistance. The greater the surface area of the parachute, the more the air resistance, and the more slowly the parachute will fall. |

| | |
|---|---|
| **Vocabulary** | • Gravity—the force that pulls an object to the ground<br>• Air resistance—a force caused by the friction of air that works against gravity and slows down objects that are falling to Earth |

| | |
|---|---|
| **Safety** | Stepping up on something to test the parachute should be done by adults only. |

**What You Need:**

☐ Items to choose from to construct the parachutes, coffee filters (large), tissue paper (both round and square), plastic squares

☐ Pieces of string the same length

☐ Tape

☐ Weights of different sizes

**How-to:**

1. Construct a parachute from the items available.

2. Lay your parachute flat on the table.

3. Tape 4 pieces of string to the edge of the parachute.

4. Attach a weight to the end of each string.

5. Give to an adult to test.

**Things to Think About:**

1. What is making the parachute fall?

2. What is slowing the parachute down?

3. How can you modify the parachute to make it fall more slowly?

4. How are parachutes helpful?

**Instructions and General Information**

**Estimated Activity Time:** 10 minutes

| Teacher Tips: PLANT DISSECTION | |
|---|---|
| **Core Content** | Structures and Processes |
| **Crosscutting Concepts** | Structure and Function |
| **Science and Engineering Practices** | Analyzing and Interpreting Data |

| | |
|---|---|
| **Guiding Questions** | What are the parts of a flower?<br>What are the parts of a seed? |
| **BEFORE the Event** | Alstroemeria flowers work well for this activity because of their shape. Make sure to purchase them in advance so that they are fully open for the event. Sometimes florists are willing to donate flowers that are too old to sell. Soak the lima bean seeds the night before the event so they are easy to open. |
| **DURING the Event** | To make it easier to manipulate the flowers, put a piece of tape on an index card sticky side up. Lay the cut flower on the tape. Keep the lima beans in the water until you are ready to use them so they stay soft and easy to take apart. Providing students with diagrams of the parts of a flower and seed will help them learn the terms for the parts and give them something to take with them for future reference. |
| **AFTER the Event** | Sweep floors to remove all flower and seed parts. |
| **Additional Resources** | Provide models of flowers for the students to look at while they are waiting their turn to dissect. Also provide posters that show different ways seeds are dispersed. |
| **Things to Think About** | 1. What are the parts of the flower? Did you find them in your dissection? (Petal, sepal, stamen, and pistil.)<br><br>2. Plants need to be pollinated. This means pollen is carried from one plant to another. Can you think of some ways this can happen? (Bees, butterflies, bats, birds, wind.)<br><br>3. Can you identify the parts of a seed? Did you find them all? (Seed coat, food storage, embryo.)<br><br>4. What are some ways that seeds can be dispersed? (Animals, humans, wind, water, seed shape [burr, wing, parachute], gravity, bursting.) |

## Activity: PLANT DISSECTION

### Program Information Section
*List location and hosts' names*

| | |
|---|---|
| **Science Behind the Activity** | Just like people, plants have specialized parts that help them survive and produce offspring. For plants to reproduce, pollen, which is produced by plants, must be carried from one flower to another to produce seeds. |

**Parts of a Flower**
Petal
Stamen
Pistil
Leaf
Sepal
Stem

**Parts of a Seed**
Embryo (baby plant)
Food Storage
Seed Coat

| | |
|---|---|
| **Vocabulary** | • Flower—the part of the plant that attracts pollinators to help the plant reproduce and make seeds to grow new plants<br>• Seed—the part of the plant that contains a young plant<br>• Pollen—a substance produced by the plant that forms new seeds<br>• Pollinators—organisms that carry pollen from one plant to another |
| **Safety** | Watch for allergies to plants and pollen. Be sure to have students wash their hands after this activity. |

**What You Need:**

☐ Lima bean seeds and seed diagram
☐ Flowers and flower diagram
☐ Magnifying glasses
☐ Index cards
☐ Tweezers

**How-to:**

Station 1: Flower

1. Cut the flower in half.
2. Lay the flower, cut side up, on a piece of tape.
3. Use tweezers to open the flower.
4. Identify the parts using the diagram.

Station 2: Seed

1. Gently pull the seed coat off.
2. Open the food storage.
3. Find the embryo (baby plant) inside.

**Instructions and General Information**

**Things to Think About:**

1. What are the parts of the flower? Did you find them in your dissection?
2. Plants need to be pollinated. This means pollen is carried from one plant to another. Can you think of some ways this can happen?
3. Can you identify the parts of a seed? Did you find them all?
4. What are some ways that seeds can be dispersed?

**Estimated Activity Time:** 10+ minutes

| Teacher Tips: ROLLER COASTER CHALLENGE | |
|---|---|
| **Core Content** | Force and Motion |
| **Crosscutting Concepts** | Energy and Matter |
| **Science and Engineering Practices** | Developing and Using Models |

| | |
|---|---|
| **Guiding Question** | How can you design a roller coaster track that keeps the marble moving from beginning to end, without it falling off? |
| **BEFORE the Event** | Plan to have this event in a big open space. A gym or a room with little or no furniture will work well. Cut the insulation tubing ahead of time (lengthwise). Keep these cut pieces of tubing in a large box to reuse over and over again. |
| **DURING the Event** | Tape one end of the tubing to the wall so that participants have free hands to make loops. Sections can be taped together with duct tape to make for longer tracks. It makes it easier to tape the cup at the end to catch the marble. You might want to keep a record of times to see who can make the fastest roller coaster. |
| **AFTER the Event** | Make sure all marbles are accounted for. |
| **Additional Resources** | Have pictures of roller coasters with diagrams explaining the potential and kinetic energy. Decorate the room like the nearest theme park. |
| **Things to Think About** | 1. How does the height of the track affect the marble? (The higher the start, the more potential energy it has and the faster it will go, until the direction is changed.)<br>2. How many loops can you add and still get the marble to make it to the end? (This depends on the angle of the track and the momentum of the marble.)<br>3. What would you have to change to add another track but still make sure the marble makes it to the end? (The starting point of the track may have to be placed higher in order to keep it at a slope so the marble keeps going.)<br>4. At what point does your marble have the greatest potential energy? (Just before you release it, at the start of the track.) |

## Activity: ROLLER COASTER CHALLENGE

### Program Information Section
*List location and hosts' names*

| | |
|---|---|
| **Science Behind the Activity** | Gravity and inertia are the forces that make a roller coaster so much fun. Gravity pulls the car down, while inertia moves it forward. Just before a roller coaster ride begins, it has potential energy. When it starts moving, the potential energy is transferred to kinetic energy. Each time a roller coaster goes over a hill, it loses energy, so each loop must be lower than the one before. |

High Potential Energy
Low Kinetic Energy

Low Potential Energy
High Kinetic Energy

| | |
|---|---|
| **Vocabulary** | • Gravity—the downward force that pulls objects to the Earth<br>• Inertia—the tendency of an object in motion to keep moving<br>• Potential energy—the energy stored in an object<br>• Kinetic energy—energy of motion |
| **Safety** | Watch small children with the marbles. They should not go in children's mouths, noses, or ears. Make sure they aren't on the floor to slip on. |

**What You Need:**

☐ 6-foot sections of pipe insulation tubing, precut lengthwise
☐ Marbles
☐ Duct tape
☐ Stopwatch

**How-to:**

1. Design a roller coaster track using the pieces of pipe insulation. Add hills and loops to your track to make it exciting.
2. Put a marble at the top and release.
3. Redesign your track until the marble rolls the entire way without falling off.
4. Time your marble on your track. Can you redesign your track so it is faster?

**Things to Think About:**

1. How does the height of the track affect the marble?
2. How many loops can you add and still get the marble to make it to the end?
3. What would you have to change to add another track but still make sure the marble makes it to the end?
4. At what point does your marble have the greatest potential energy?

**Instructions and General Information**

**Estimated Activity Time:** 10 minutes

| Teacher Tips: STATIC SENSATIONS | |
|---|---|
| **Core Content** | Conservation of Energy and Energy Transfer |
| **Crosscutting Concepts** | Energy and Matter |
| **Science and Engineering Practices** | Planning and Carrying Out Investigations |

| | |
|---|---|
| **Guiding Questions** | What is static electricity? <br><br> What causes static electricity? |
| **BEFORE the Event** | This station has a number of activities to explore static electricity, so use a room with lots of space and multiple tables. This works best in an open room with carpet. At one station, set up two balloons, close to each other, suspended from the ceiling on string. Make sure you have plenty of combs to give out, as they should not be reused. Have some latex-free balloons for those with allergies to latex. |
| **DURING the Event** | We add a plasma ball and static stickers for students to play with. These are the most popular items at this activity! Monitor students so that they don't trip while exploring. |
| **AFTER the Event** | Make sure room is reset and all the balloons are collected. Sweep up any salt and pepper or confetti. |
| **Additional Resources** | Because lightning is a form of static electricity, you might want to include décor to make the room look like Dr. Frankenstein's lab! |
| **Things to Think About** | 1. What did you have to do to make your hair attract to your balloon? (The balloon has to be "charged" with the wool. This moves electrons so that they build up on the balloon. The charged balloon is attracted to objects with the opposite charge.) <br><br> 2. What happens to the pepper and confetti in Station 2? Why? (The charged comb attracts objects with the opposite charge.) <br><br> 3. What happens to the balloons in Station 3? Why? (The charged balloon repels objects with a like charge.) |

National Science Teachers Association

## Activity: STATIC SENSATIONS

**Program Information Section**
*List location and hosts' names*

| | |
|---|---|
| **Science Behind the Activity** | Atoms are made up of smaller parts: protons (+), electrons (–), and neutrons, which are neutral. When atoms gain or lose electrons, they become charged. Static electricity is the result of charges building up on the surface of an object because the electric charges are not balanced. Like charges will repel, or push objects away from each other, while opposite charges will attract. These charges can build up on a surface of an object, and when they meet, cause a shock! |
| **Vocabulary** | • Electric charge—a characteristic of matter based on the electric charge. This is dependent on the balance of protons and electrons. Matter can only gain or lose electrons, not protons. So when an object gains electrons, it has a negative charge, and when it loses electrons, it will have a positive charge.<br>• Static electricity—a charge built up on the surface of an object from an imbalance of electrons. Because it stays in one place, it is "static." |
| **Safety** | Be aware of latex allergies. Provide latex-free balloons to students who are allergic to latex. NEVER let students share or reuse combs. |

**What You Need:**

☐ Balloons
☐ Combs (light colored)
☐ Mirror

☐ Tissue paper confetti
☐ Salt and pepper mixture in a bowl
☐ Pieces of wool (or wool scarves)

**How-to:**

Station 1:

1. Blow up a balloon.
2. Rub it with a piece of wool to charge it.
3. Place it on your hair and look in the mirror!
4. Put it on the wall and let it go.

Station 2:

1. Rub the comb with wool to charge it.
2. Put it over the bowl of salt and pepper.
3. Try to pick up pieces of confetti with your comb.

Station 3:

1. Rub the hanging balloons with pieces of wool.
2. What happens to the balloons when you let them go?

**Instructions and General Information**

**Things to Think About:**

1. What did you have to do to make your hair attract to your balloon?
2. What happens to the pepper and confetti in Station 2? Why?
3. What happens to the balloons in Station 3? Why?

**Estimated Activity Time:** 10+ minutes

| Teacher Tips: STRAW ROCKETS | |
|---|---|
| Core Content | Motion and Stability: Forces and Interactions |
| Crosscutting Concepts | Structure and Function |
| Science and Engineering Practices | Constructing Explanations and Designing Solutions |

| | |
|---|---|
| **Guiding Question** | How can you design a straw rocket to go the farthest distance? |
| **BEFORE the Event** | Make sure to have a launching area that is away from the traffic flow. Mark the area with tape or cones. It makes it more fun to have a target or bucket at the end of the track. |
| **DURING the Event** | Encourage students to use the engineering cycle to design, test, and then improve their rocket. |
| **AFTER the Event** | Make sure all the furniture is back in place around the room and all straws are accounted for. |
| **Things to Think About** | 1. Which forces are at work to make your straw rocket fly better? (Lift and thrust.)<br><br>2. Which forces are working against your rocket? (Drag and weight [gravity].)<br><br>3. When was the force of thrust applied to your rocket? (When it was launched by blowing through the small straw.)<br><br>4. What changes did you make to your rocket? Why did you make them? (Discuss with students how their designs might be improved to add lift or thrust, or may reduce drag [friction] or weight [the pull of gravity]).<br><br>5. How did the changes affect your rocket? (Answers may vary.) |

## Activity: STRAW ROCKETS

### Program Information Section
*List location and hosts' names*

| | |
|---|---|
| **Science Behind the Activity** | When launching straw rockets air pressure provides thrust and lift. Other forces, weight (gravity) and drag, work to slow it down. The length of a rocket will change its flight distance—a longer straw is filled with more air and provides more thrust (pushes on the straw for a longer time) causing it to fly farther. Fins and nose weight can improve a rocket's stability. |

**Forces of Flight**

Lift, Thrust, U.S.A., Drag, Weight

| | |
|---|---|
| **Vocabulary** | • Force—a push or a pull<br>• Thrust—the forward force on a rocket<br>• Drag—a force caused by friction of the air pushing on the rocket, which holds back the rocket and acts against thrust<br>• Lift—the air moving over and under the rocket, which holds it up<br>• Weight—a measure of the force of gravity pulling the rocket down |
| **Safety** | Participants must wear eye protection. Make sure test area is set up away from the design space. Use caution when working with scissors, which can cut or puncture skin. Immediately clean up any clay at the end of this activity. Dry clay contains silica, which is a respiratory hazard. Participants should wash hands with soap and water upon completing this activity. |

**What You Need:**

☐ Paper        ☐ Permanent markers        ☐ Clay
☐ Tape         ☐ Straws (two sizes, one that        ☐ Safety goggles
☐ Scissors         can fit into the other)

**How-to:**

1. Put a small piece of clay in the end of the larger straw.

2. Add fins to your large straw and decorate it using the markers.

3. Put the large straw over the small straw.

4. Aim and blow through the small straw to launch your rocket.

**Things to Think About:**

1. Which forces are at work to make your straw rocket fly better?

2. Which forces are working against your rocket?

3. When was the force of thrust applied to your rocket?

4. What changes did you make to your rocket? Why did you make them?

5. How did the changes affect your rocket?

**Instructions and General Information**

**Estimated Activity Time:** 10 minutes

| Teacher Tips: SURFACE TENSION | |
|---|---|
| **Core Content** | Structure and Properties of Matter |
| **Crosscutting Concepts** | Energy and Matter |
| **Science and Engineering Practices** | Planning and Carrying Out Investigations |

| | |
|---|---|
| **Guiding Questions** | How does surface tension affect movement in liquids?<br>How many drops of water can a penny hold before the water slides off? |
| **BEFORE the Event** | While you can use any shallow dish for these activities, we use petri dishes. Have at least 20 on hand so that dishes can be washed and dried between uses. This activity is set up in stations. Make sure you are in a room with a sink. To prepare for the water strider activity, precut 2 in. pieces of old holiday light wires and strip both ends. Leave a ½ in. piece of plastic in the center. We used pieces of old chip bags for the bugs, rather than buying cellophane. For the coins, have a number of pennies as well as nickels, dimes and quarters. For the tye-dye milk, use WHOLE milk only. Have buckets by each table to pour the used milk into. Cover the tables with a tablecloth. These activities are messy, so have extra table covers to change out during the event. |
| **DURING the Event** | The water striders only float if the "legs" are fanned out just right. If it takes more than one try to get the strider to float, dry it off between attempts. For the coin activity, it is helpful to have younger students practice squeezing the pipette and count drops into the cup before starting on the penny. Make sure to dry coins before drops are added. To make it more fun, we added a chart where participants could write how many drops their pennies held. |
| **AFTER the Event** | Pass the chart on to a teacher who can use the chart in his or her class to work with data. Be sure to clean up the tables and floor. Dispose of the used milk in a toilet rather than pouring it in water fountains. |
| **Additional Resources** | Provide pictures of insects sitting on top of the water and a swimmer with the water flowing over his or her head. You may also wish to have Isopropyl alcohol at Station 2 for students to compare liquids. |
| **Things to Think About** | 1. What keeps your water strider from sinking? (The surface tension of the water.)<br>2. What coin holds the most drops? Why? (The bigger the coin, the more surface area it has and thus the more water it will hold.)<br>3. What happens to the food coloring in the milk? (It does not move much because the milk fat has high surface tension.)<br>4. What happens when a drop of soap is added? (The soap is breaking the bonds in the fat.) |

## Activity: SURFACE TENSION

### Program Information Section
*List location and hosts' names*

| | |
|---|---|
| **Science Behind the Activity** | Water has an elastic "skin" where the molecules are attracted to each other. This is called surface tension. Because of surface tension, water striders and some objects can bend the surface of water without breaking it. When you put drops of water on a penny, the water molecules hold on to each other and create a dome of water on it. Soap breaks the attraction of molecules (cohesion) and reduces the surface tension. When soap is put into milk, it tries to bond with the fat and breaks it apart, and the food dye dances around! |

**Surface Tension: Water's cohesion is strongest at the surface because it can not attract in all directions**

| | |
|---|---|
| **Vocabulary** | • Cohesion—the nonbonding attraction of molecules<br>• Surface tension—the attraction of molecules at the surface of a liquid |
| **Safety** | Students should wear goggles when working with food coloring and soap. Students should not drink any liquid used in this activity. Keep floors dry. |

**What You Need:**

Station 1:
- ☐ Small dish of water
- ☐ 2 in. pieces of wire from lights, with ends stripped
- ☐ 1 in. pieces of cellophane
- ☐ Permanent markers

Station 2:
- ☐ Coins (pennies)
- ☐ Pipette
- ☐ Paper towels
- ☐ Paper plates (small)
- ☐ Water

Station 3:
- ☐ Petri dish
- ☐ Food coloring
- ☐ Milk (whole)
- ☐ Dish soap
- ☐ Cotton swabs
- ☐ Safety goggles

**How-to:**

1. Station 1: Make your own water strider! Create a bug body out of pieces of cellophane. Tape it to the plastic wire covering in the middle. Fan out the wire "legs" of the bug. See if you can get it to float in water!

2. Station 2: First, practice using the dropper to make the drops about the same size. Then count the drops your penny can hold before the water slides off. See who can get the most water on a penny!

3. Station 3: Make tie-dye milk! Pour a thin layer of milk into the dish. Add several drops of food coloring. Dip a cotton swab into the dish, touch the milk, and watch!

**Things to Think About:**

1. What keeps your water strider from sinking?
2. What coin holds the most drops? Why?
3. What happens to the food coloring in the milk?
4. What happens when a drop of soap is added?

**Instructions and General Information**

**Estimated Activity Time:** 10 minutes

# Chapter 12

# Advanced-Level Activities

The activities we present in this final chapter are those that we think are at higher conceptual levels. As before, you can choose a variety of activities, selecting for both levels and concepts. We hope that you will find this book useful as you plan your Family Science Night events. The ideas we've shared throughout the book are based on more than a decade of experiences at different levels. We have truly found these events to be the highlight of our careers, especially by putting students in charge. You will find, as we have, that your events will grow and evolve in unexpected ways. You'll have ideas we haven't thought of and you'll find new ways to engage your school and community in hands-on science. The activities in this final chapter are as follows:

- Balloon-Powered Cars—build a race car for this engineering challenge
- Cartesian Divers—use air pressure to make a pipette sink and float
- Chemical Reactions—find out what happens when you mix baking soda and vinegar
- Hall Rollers—convert elastic potential energy to kinetic energy with this device
- Hovercraft—build your own hovercraft from a used CD
- Krazy Kaleidoscopes—explore reflection by making this popular toy
- Mighty Lungs—discover how your lungs work in this activity
- Modeling the Rock Cycle—explore the transformations of the rock cycle with crayons
- Owl Pellets—explore an owl pellet to learn about energy and food chains
- Paper Airplanes—build a paper airplane in this engineering challenge
- Rainbow Stacking—explore the concept of density by building your own column
- Work It, Circuits!—find out how electricity flows in circuits in this activity

| Teacher Tips: BALLOON-POWERED CARS | |
| --- | --- |
| **Core Content** | Forces and Motion |
| **Crosscutting Concepts** | Cause and Effect |
| **Science and Engineering Practices** | Constructing Explanations and Designing Solutions |

| | |
| --- | --- |
| **Guiding Question** | What design will help your car move the fastest and farthest? |
| **BEFORE the Event** | A number of materials can be used for this engineering challenge. The ones we've suggested are based on our experience, but other materials can be used. Set this activity up where students can spread out and experiment. This is a popular activity and can get messy. Using an assembly line for materials can help keep this station organized. |
| **DURING the Event** | Have an area set aside where the participants can test their cars. You might want to encourage students to time their trials or measure distance. Don't forget to encourage them to use the engineering cycle to modify their designs to improve them. |
| **AFTER the Event** | Make sure area is cleaned up and all balloons are cleared out of the room. |
| **Additional Resources** | You can offer different kinds of wheels—for example, lids to cups from fast-food restaurants. Include some stopwatches for timing and meter sticks for measuring distance. Decorate the room like a race track, and announce the "drivers" when they test their designs! |
| **Things to Think About** | 1. What makes the car move? (Air pushing out of the straw provides the force to make the car go.)<br>2. Which direction does the air come out of the balloon and which direction does your car move? (The air goes out the back and pushes the car forward.)<br>3. Does the amount of air in your balloon affect how far your car goes? Which of Newton's laws addresses this? (Newton's second law suggests that objects that have a greater force applied, will move more. Encourage students to experiment with this variable to determine that more air provides more energy and can cause the car to move a greater distance, doing more "work.")<br>4. How can you modify your car to make it go in a circle? (If the front axle is taped at an angle, the car will go in a circle.) |

## Activity: BALLOON-POWERED CARS

### Program Information Section
*List location and hosts' names*

| | |
|---|---|
| **Science Behind the Activity** | Newton's laws tell us how objects move. The first law states that objects don't move until a force acts on them. The second law tells us that the more force that is applied, the greater the movement. Newton's third law of motion states that forces act in pairs. When a force is applied to an object, an equal and opposite reaction occurs. In your balloon car, the air pushes out the back and the car goes forward. When your car is moved by a force, energy is transferred and work is done. |

| | |
|---|---|
| **Vocabulary** | • Force—a push or pull that changes the motion of an object<br>• Energy—the ability to cause change<br>• Work—what happens when an object is moved by a force |
| **Safety** | To avoid allergic reactions to latex, use latex-free balloons. |

**What You Need:**

☐ Balloon—your engine
☐ 3 straws—1 for releasing air from the balloon and 2 for axles
☐ 4 Life Saver candies—for wheels
☐ Alternate wheels (lids or paper disks)
☐ Hole punch—to create a hole for the straw "axles" to go through
☐ Tape—to secure the straw to the balloon
☐ A car "body"—e.g., a hot dog holder, plastic tray, cone, water bottle, etc.

**How-to:**

Be an engineer! Use the materials to design a car that is powered by a balloon.

1. Choose a "body" for your car and attach wheels.

2. Tape the neck of the balloon to a straw.

3. Use tape to attach the straw to the car.

4. Blow up the balloon through the straw and let it go!

5. Think about how to improve your design. Try it out!

**Things to Think About:**

1. What makes the car move?

2. Which direction does the air come out of the balloon and which direction does your car move?

3. Does the amount of air in your balloon affect how far your car goes? Which of Newton's laws addresses this?

4. How can you modify your car to make it go in a circle?

**Instructions and General Information**

**Estimated Activity Time:** 10+ minutes

# Chapter 12

| Teacher Tips: CARTESIAN DIVERS | |
|---|---|
| **Core Content** | Motion and Stability: Forces and Interactions |
| **Crosscutting Concepts** | Cause and Effect |
| **Science and Engineering Practices** | Developing and Using Models |

| | |
|---|---|
| **Guiding Question** | How does a Cartesian diver work? |
| **BEFORE the Event** | Peel the paper off the water bottles. For the diver to work, you have to have a bubble the right size in the pipette. If the bubble is too big, it will not sink or it will be too hard to sink with kid pressure applied. If the bubble is too small, it will sink automatically with no pressure. Set up some water tubs for the hosts to practice. Arrange the room with three tables for students to make their Cartesian divers in an assembly line process. Make sure the room this activity is in has a sink. Have extra Cartesian diver bottles of different sizes premade for students to play with and to occupy those that are waiting their turn. |
| **DURING the Event** | We found it helpful to set up this activity like an assembly line. The first table is an exploration area for students to play with premade divers. The second table is for cutting the pipettes and putting on the hex nuts. At the third table, help students put their divers in the water tubs to practice getting the right size bubbles in the pipettes before putting them into their own bottles. |
| **AFTER the Event** | Make sure all tables are dry and sink is clean. Pick up all cut pipette pieces; they look almost invisible on the floor. |
| **Additional Resources** | It would be great if you knew someone with scuba gear who could talk to those waiting and explain why the pressure is important as a diver. The deeper you go in water, the smaller the bubble in your tank, and the less air you have to breathe. |
| **Things to Think About** | 1. What happens when you squeeze the bottle? (When you squeeze, the water pushes the air up into the pipette, compressing it. It becomes denser and the diver sinks.)<br>2. What happens when you stop squeezing the bottle? (When you release the bottle, the air in the pipette expands and the diver rises.) |

National Science Teachers Association

## Activity: CARTESIAN DIVERS

### Program Information Section

*List location and hosts' names*

| | |
|---|---|
| **Science Behind the Activity** | Cartesian divers rise and sink because of the changing air pressure in the pipette. When the bottle is squeezed, the water pushes the air in the pipette and compresses it. It becomes less buoyant and sinks. When the bottle is released, the compressed air expands and forces water out of the diver, reducing density of the air bubble and allowing it to float to the top of the bottle. |

Squeezing the bottle makes the air compress (become more dense) and the diver sinks. Releasing the bottle lets the air expand (become less dense) and the diver rises.

| | |
|---|---|
| **Vocabulary** | • Air pressure—the weight of the air molecules pressing down<br>• Compress—to squeeze things together<br>• Buoyancy—the tendency of an object to float |
| **Safety** | If using recycled bottles, be sure to sterilize the bottles. |

**What You Need:**

☐ Water bottles

☐ Hex nuts

☐ Pipettes

**How-to:**

1. Cut off the tip of pipette.

2. Screw a hex nut onto the tip, close to the bulb.

3. Put just enough water in the pipette bulb to float on top of the water.

4. Put the diver in a filled water bottle and screw top on bottle (tight).

5. Gently squeeze water bottle and watch your diver go down.

**Things to Think About:**

1. What happens when you squeeze the bottle?

2. What happens when you stop squeezing the bottle?

**Instructions and General Information**

**Estimated Activity Time:** 10+ minutes

| Teacher Tips: CHEMICAL REACTIONS | |
|---|---|
| Core Content | Matter and Its Interactions |
| Crosscutting Concepts | Energy and Matter |
| Science and Engineering Practices | Planning and Carrying Out Investigations |

| | |
|---|---|
| **Guiding Questions** | What is a chemical reaction?<br><br>How can you tell when a chemical reaction occurs? |
| **BEFORE the Event** | We like to use recycled 8 oz. water bottles from the school cafeteria for this activity. Be sure to sterilize the mouths of the bottles. Bottle caps are not needed. For this activity, be sure that your tables are covered. Prefill the water bottles with about ¼ cup vinegar before the event. Put baking soda in small cups on tables. |
| **DURING the Event** | Make sure students put on goggles first! Have students work in pairs. One should hold open the balloon over a paper plate, while the other student puts about 3 teaspoons of baking soda into the balloon. You might want to have some funnels to make this easier and reduce the mess. |
| **AFTER the Event** | Make sure that the tables and floor are cleaned. |
| **Additional Resources** | Put out a pan with baking soda and cups with vinegar. Let students use pipettes to put vinegar on the baking soda to watch the reaction directly. You can color the vinegar if you choose. You might also want to include some pieces of limestone rock for students to observe that the same reaction happens in nature when vinegar is put on the rock. Additionally, provide some pH paper for students to see the difference between vinegar as an acid and baking soda as a base. |
| **Things to Think About** | 1. What happened when you combined the baking soda and vinegar? (The substances should fizz and bubble and the balloon blow up.)<br><br>2. How can you tell a chemical reaction occurred? (A gas is produced—in this case, carbon dioxide gas, which fills up the balloon.)<br><br>3. Did the weight of your bottle change during the reaction? (The weight should remain the same during this experiment. Talk to students about the law of conservation of mass, which tells us that even though new substances are being formed, the mass stays the same.) |

## Activity: CHEMICAL REACTIONS

### Program Information Section
*List location and hosts' names*

| | |
|---|---|
| **Science Behind the Activity** | Mixing baking soda and vinegar produces an acid-base reaction. Baking soda is the base and vinegar is an acid. When these two substances are mixed, they react and form new substances (sodium acetate, water, and carbon dioxide gas). There are several ways to know if a chemical reaction occurs; one is by producing a gas, which can be observed in this reaction as the balloon blows up. Even without the balloon, the fizzing and bubbling that happen are a good indication that a gas is being produced. Other ways to know when a chemical reaction occurs is the production of heat, a change of color, and the formation of a solid. |

Baking soda in balloon

Vinegar

| | |
|---|---|
| **Vocabulary** | • Chemical reaction—a process where mixing two substances together causes a reaction and produces new substances<br>• Acid—a chemical compound that has a lot of hydrogen ions (acids are rated between 0 and 7 on the pH scale)<br>• Base—a chemical compound that has a lot of hydroxide ions (bases are rated between 7 and 14 on the pH scale) |
| **Safety** | Participants MUST wear goggles and nonlatex aprons in this activity. If using recycled bottles, make sure they are sterilized before use. Students should wash their hands with soap and water after completing this activity. Have a portable eye wash bottle available for use in case of a chemical splash. |

**What You Need:**

☐ Balloons  ☐ Vinegar  ☐ Kitchen scales  ☐ Aprons
☐ Baking soda  ☐ 8 oz. plastic bottles (empty)  ☐ Safety goggles

**How-to:**

1. Put approximately 3 teaspoons of baking soda in the balloon.
2. Put ¼ cup vinegar in the empty water bottle.
3. Place the balloon over the mouth of the bottle, keeping substances separate.
4. Weigh the bottle of vinegar with the balloon attached on the scale.
5. Lift the balloon to pour the baking soda into the balloon.
6. Watch what happens!
7. Weigh the bottle again.

**Things to Think About:**

1. What happened when you combined the baking soda and vinegar?
2. How can you tell a chemical reaction occurred?
3. Did the weight of your bottle change during the reaction?

**Instructions and General Information**

**Estimated Activity Time:** 10+ minutes

| Teacher Tips: HALL ROLLERS | |
|---|---|
| **Core Content** | Conservation of Energy and Energy Transfer |
| **Crosscutting Concepts** | Energy and Matter |
| **Science and Engineering Practices** | Planning and Carrying Out Investigations |

| | |
|---|---|
| **Guiding Questions** | How is potential energy stored?<br>How is potential energy converted to kinetic energy? |
| **BEFORE the Event** | Different size cup lids will work for this activity. You can buy them in bulk at wholesale clubs, or you can ask local restaurants or your lunchroom manager to donate them. To prepare, prepunch holes in the middle of the lids. Precut the card stock to the right size for creating a cylinder. You may want to test out different sizes. Set up a long table available for testing the hall rollers. |
| **DURING the Event** | One activity host can help three or four students at a time. Watch for beads that fall on the floor and can become a trip hazard. |
| **AFTER the Event** | Make sure all beads are picked up off the floor. |
| **Additional Resources** | You can purchase commercial hall rollers for students to work with. Set up a center with other toys that store elastic potential energy, such as "chattering teeth," rubber band cars, and wind-up toys. Include some meter sticks for measuring the distance hall rollers travel. Set up the room with a racing theme! |
| **Things to Think About** | 1. Where did the energy come from that was "stored" in the rubber band? (From winding it up.)<br>2. What happens if you wind the rubber band more? (When the rubber band is wound tighter, it will spin faster. If it is too tight, it pulls too hard on the wheels, and they will not turn or the roller may collapse.)<br>3. How is energy "conserved" in this activity? (It is never created; it only changes form, from kinetic to potential to kinetic.)<br>4. Can you change the direction of the roller? (Change how much the straw extends out and see if it changes the direction.) |

## Activity: HALL ROLLERS

### Program Information Section
*List location and hosts' names*

| | |
|---|---|
| **Science Behind the Activity** | Energy cannot be created or destroyed—it can only change form! Potential energy is stored and kinetic energy is in motion. When you wind up your hall roller, you provide kinetic energy that is stored as potential energy, which converts back to kinetic energy when the hall roller is released. |

Rubber band goes through the lid, then the bead, and loops around the straw.

Rubber band goes through the paper clip here. Tape the paper clip to the lid.

Roll the cardstock and tape. Attach a plastic lid to each end.

| | |
|---|---|
| **Vocabulary** | • Potential energy—the stored energy in an object<br>• Kinetic energy—the observable movement of an object<br>• Law of conservation of energy—the idea that energy cannot be created or destroyed (it can only change form) |
| **Safety** | Make sure beads are immediately picked up off the floor, to prevent someone from slipping. Watch to make sure small children don't put beads in ears, mouths, or noses. Use caution when working with scissors, which can cut or puncture skin. Eye protection is required for this activity. |

**What You Need:**

☐ Rubber bands        ☐ Paper clips        ☐ Plastic          ☐ Colored pencils
☐ Coffee stirrers     ☐ Tape                 drink lids       ☐ Scissors
   or straws          ☐ Beads              ☐ Cardstock        ☐ Safety goggles

**How-to:**

1. Punch holes in the center of two lids, if not already done.
2. Decorate your cardstock with the pencils.
3. Roll the cardstock to form a cylinder. Attach a plastic lid on each end.
4. Run a rubber band through the cylinder and out the hole in each lid.
5. On one end, attach a paper clip to the rubber band then tape it down.
6. On the other end, run the rubber band through the bead and insert a straw through the loop that is created at the end of the rubber band.
7. Spin the straw around in circles (kinetic energy) several times until the rubber band is properly wound (potential energy).
8. Set your hall roller down and release it. Watch it go (kinetic energy)! ☺

**Things to Think About:**

1. Where did the energy come from that was "stored" in the rubber band?
2. What happens if you wind the rubber band more?
3. How is energy "conserved" in this activity?
4. Can you change the direction of the roller?

**Instructions and General Information**

**Estimated Activity Time:** 10+ minutes

| Teacher Tips: HOVERCRAFT | |
|---|---|
| Core Content | Forces and Motion |
| Crosscutting Concepts | Cause and Effect |
| Science and Engineering Practices | Developing and Using Models |

| | |
|---|---|
| **Guiding Question** | How does a hovercraft work? |
| **BEFORE the Event** | We use film canisters, but you can also use bottle caps (or pop-top liquid soap bottle caps) instead to save money. Predrill a hole in the bottom of the film canisters or bottle caps. Preglue the film canisters or bottle caps to the CDs before the event. We suggest hot glue. Make sure to be in a room that has long tables to test the hovercraft. |
| **DURING the Event** | Keep tables clean. Dirty tables will keep the hovercraft from performing well. It takes two people to get the hovercraft moving. One person needs to pinch the balloon to keep the air in, while the other person stretches it over the bottle cap. To control the flow of a large crowd, set up a table where the students can decorate the CD part of the hovercraft with permanent markers while waiting. |
| **AFTER the Event** | Make sure the area is cleaned up and tables are returned to their original position. |
| **Additional Resources** | Provide pictures or videos of large hovercrafts. You might also want to have a couple of purchased hovercraft toys for students to play with. If you know of high school or college students who have made large hovercrafts, invite them out to demonstrate at this activity. |
| **Things to Think About** | 1. How does a hovercraft work? (The air released through the holes creates a cushion of air under the disc and friction is reduced.)<br>2. Does the amount of air in the balloon change how the hovercraft works? (More air in the balloon provides more volume and will make the hovercraft move for a longer period of time.)<br>3. Does the type of surface affect how fast or far it can move? (The smoother the surface, the less friction it will have and the faster it can move.) |

## Activity: HOVERCRAFT

### Program Information Section
*List location and hosts' names*

| | |
|---|---|
| **Science Behind the Activity** | When objects move, friction between surfaces works against motion and slows them down. If you can reduce friction, things move easier. Hovercrafts reduce friction by moving on a thin layer of air. In this case, the air from the balloon moves through the holes and rushes outward in all directions under the disc, creating a cushion of air that allows the disc to hover freely over the surface. This cushion of air reduces the friction and allows the hovercraft to glide easily across the table when given a push. |

**Drill hole in bottle cap. Glue bottle cap onto center of CD with hole down and open end up.**

| | |
|---|---|
| **Vocabulary** | • Friction—a force that resists motion. It is caused by objects rubbing against each other or by air when an object is moving through it. |
| **Safety** | To avoid allergic reactions to latex, use latex-free balloons. |

**What You Need:**

☐ CDs

☐ Film canisters (no lid) or bottle cap, with small hole drilled in bottom

☐ Balloons (9 in. or 12 in.)

☐ 2 in. strips of paper

**How-to:**

1. Decorate your hovercraft (CD) with markers.

2. Roll paper into a cylinder to form a collar for the balloon.

3. Blow up balloon.

4. Twist balloon to hold in air, and thread it through the collar.

5. Attach to the top of the CD.

6. Put it down on a table, give it a nudge, and watch it GO!

**Things to Think About:**

1. How does a hovercraft work?

2. Does the amount of air in the balloon change how the hovercraft works?

3. Does the type of surface affect how fast or far it can move?

**Instructions and General Information**

**Estimated Activity Time:** 10 minutes

## Chapter 12

| Teacher Tips: KRAZY KALEIDOSCOPES | |
|---|---|
| Core Content | Light |
| Crosscutting Concepts | Structure and Function |
| Science and Engineering Practices | Constructing Explanations and Designing Solutions |

| | |
|---|---|
| **Guiding Question** | How is light reflected? |
| **BEFORE the Event** | You can purchase sheets of mirror adhesive or adhesive mylar. Precut it into rectangles 4.5 in. × 6 in. that can be folded into triangles. You might want to prefold the triangles. You can collect recycled toilet paper tubes or purchase them. Pour beads into a bowl for easy access. |
| **DURING the Event** | Small children might not have the fine motor skills to assemble their kaleidoscopes and may need help. ![Toilet Paper Tube, Foil Tent (3 sides), Beads] |
| **AFTER the Event** | Sweep up the mess, making sure no beads have escaped as they can be slipped on. |
| **Additional Resources** | Have other resources to experiment with light set up. For example, a station with a pet laser light and mirrors for students to explore will help them with understanding how light travels and how mirrors reflect light. |
| **Things to Think About** | 1. How does light travel? (In straight lines.)<br>2. Why do you see a complicated pattern of colors instead of the beads in the end cap? (Because the light is reflected off the mirrors inside.)<br>3. What do you think would happen if you had more mirrors inside your kaleidoscope? (The light would be reflected more, and you'd see more complicated patterns.) |

National Science Teachers Association

# Advanced-Level Activities

| Activity: KRAZY KALEIDOSCOPES | |
|---|---|

**Program Information Section**
*List location and hosts' names*

| **Science Behind the Activity** | Light travels in straight lines. In a kaleidoscope, you see light that is reflected through the beads and off the different surfaces of the reflective material to create beautiful patterns. As you turn the kaleidoscope, the beads shift position, and because of the way light bounces off the mirrors, you never see the same pattern twice! | Cardboard Tube / Eyepiece / Mirror Triangle / Endcap (place bag of beads here) |
|---|---|---|
| **Vocabulary** | • Reflection—when light bounces off a surface | |
| **Safety** | Make sure that children do not put the beads in their mouths, ears, or noses. | |

**What You Need:**

☐ Adhesive mirror sheets (can use adhesive mylar sheets)
☐ Cardboard craft tubes (4.5 in.)
☐ Beads
☐ Index cards
☐ Plastic snack bags
☐ Clear tape
☐ Scissors
☐ Markers or crayons

**How-to:**

1. Fold a 4.5 in. × 6 in. mirror sheet into a triangular shape that will fit inside the craft tube. The reflective surface should be inside the triangle.

2. Place the mirror triangle in the cardboard tube.

3. Put some beads (approximately 1 spoonful) into a plastic snack bag and put into one end of the tube. Tape it so it doesn't fall out.

4. Cut a circle out of the index card that will fit on one end of the tube. Punch a "peep" hole in the center and tape it to the other end of the tube.

5. Decorate the outside of the tube. Hold your kaleidoscope up to the light to see amazing patterns appear!

**Things to Think About:**

1. How does light travel?

2. Why do you see a complicated pattern of colors instead of the beads in the end cap?

3. What do you think would happen if you had more mirrors inside your kaleidoscope?

**Instructions and General Information**

**Estimated Activity Time:** 10+ minutes

| Teacher Tips: MIGHTY LUNGS | |
|---|---|
| Core Content | From Molecules to Organisms: Structure and Processes |
| Crosscutting Concepts | Structure and Function |
| Science and Engineering Practices | Developing and Using Models |

| | |
|---|---|
| **Guiding Question** | How do our lungs help us breathe? |
| **BEFORE the Event** | Sterilize recycled bottles and precut the bottoms off before the event. It is difficult for younger children to cut the plastic, and the plastic can be sharp. Test out different kinds of water bottles to see which ones are durable enough to hold a cut balloon. |
| **DURING the Event** | Younger children will need assistance stretching the cut balloon over the opening of the bottle. Use tape or a rubber band on the diaphragm balloon to make sure it has a good seal and does not pull off. |
| **AFTER the Event** | Make sure all the cut pieces of balloons are cleaned up in the room. |
| **Additional Resources** | A model of a lung from a doctor's office would be helpful. You can also have a stethoscope available so students can hear their breathing. Make sure to sterilize the earbuds on the stethoscope after each use. If a nearby high school will allow you, borrow a human anatomy model to show students how the lungs fit into the body. |
| **Things to Think About** | 1. What happens when you pull the diaphragm down? (The air is pulled into the balloon [lung] in the bottle.)<br>2. What happens when you push the diaphragm up? (The air is pushed out of the balloon [lung] in the bottle.)<br>3. How is this model different from our lungs? (We have two lungs instead of one.)<br>4. Put your hands on your body, at the bottom of your rib cage. Breathe in and out deeply. Do you feel your diaphragm? |

## Activity: MIGHTY LUNGS

### Program Information Section
*List location and hosts' names*

| | | |
|---|---|---|
| **Science Behind the Activity** | Breathing requires the work of not only your lungs but also a special muscle called a diaphragm. This is located beneath your lungs at the bottom of your rib cage. When you breathe in, your diaphragm moves downward, pulling air into your lungs. When breathing out, your diaphragm pushes up, forcing the air out of your lungs. | Windpipe — Lung — Rib cage — Diaphragm |
| **Vocabulary** | • Diaphragm—a large muscle that assists in breathing<br>• Lungs—a pair of organs that take in air and pass oxygen to your blood | |
| **Safety** | Make sure recycled water bottles are sterilized. To avoid allergic reactions to latex, use latex-free balloons. | |

**What You Need:**

☐ Empty water bottles
☐ Balloons
☐ Tape or rubber bands
☐ Scissors

**How-to:**

1. Cut a balloon below the neck.
2. Stretch the cut balloon over the open end of a bottle and secure it with tape or a rubber band.
3. Push an uncut balloon in through the top of the bottle, and stretch the neck of the balloon over the mouth of the bottle.
4. Pull the cut balloon (diaphragm) and watch it fill the balloon in the bottle (lung) with air.

**Things to Think About:**

1. What happens when you pull the diaphragm down?
2. What happens when you push the diaphragm up?
3. How is this model different from our lungs?
4. Put your hands on your body, at the bottom of your rib cage. Breathe in and out deeply. Do you feel your diaphragm?

**Instructions and General Information**

**Estimated Activity Time:** 10+ minutes

| Teacher Tips: MODELING THE ROCK CYCLE | |
|---|---|
| **Core Content** | Earth's Materials and Systems |
| **Crosscutting Concepts** | Stability and Change |
| **Science and Engineering Practices** | Developing and Using Models |

| | |
|---|---|
| **Guiding Questions** | Are all rocks formed the same way? <br> How does rock change during the rock cycle? |
| **BEFORE the Event** | You can recycle old, used crayons for this activity. Set up table coverings so that it is easy to clean up crayon shavings. |
| **DURING the Event** | Have students use two different colors of crayon. We use craft sticks to "shave" the crayons; however, other items can be used (e.g., crayon sharpeners, plastic knives). Put the shavings in a snack bag to press into sedimentary rock. For the metamorphic rock, place the shavings into an aluminum cupcake tin or metal measuring cup to apply heat with the hair blower. |
| **AFTER the Event** | Make sure blow dryers are cooled down before packing them up. |
| **Additional Resources** | Have samples of igneous, metamorphic, and sedimentary rocks and a poster of the rock cycle. Also provide a magnifying glass. |
| **Things to Think About** | 1. What are the three types of rock? How do they form? (Igneous, from lava or magma; sedimentary, from sediment [grains]; metamorphic, from heat and pressure.) <br> 2. Sedimentary rocks are made from "parent" rocks. What is the parent rock for your sedimentary rock? (The crayons were the "parent" rock.) <br> 3. What do you think happens next to metamorphic rock? (It can either melt to become magma, and eventually igneous rock, or it can become weathered into sediment and become sedimentary rock.) |

## Activity: MODELING THE ROCK CYCLE

### Program Information Section
*List location and hosts' names*

| | |
|---|---|
| **Science Behind the Activity** | Rocks are constantly being recycled on Earth. They start as cooled lava or magma. These are igneous rocks. Rocks that are formed from weathered sediment are sedimentary rocks. Rocks that are buried deep underground and are changed by heat and pressure are metamorphic rocks. Rocks don't go through the rock cycle in any one order; what happens to a rock depends on the conditions it is exposed to. |

*Sedimentary — Weathering and Erosion — Heat and Pressure — Melting ("Fire Born") — Metamorphic — Igneous*

**Rock Cycle**

| | |
|---|---|
| **Vocabulary** | • Igneous—rock formed from magma or lava<br>• Sedimentary—rock formed from the products of weathering<br>• Metamorphic—rock formed from heat and pressure |
| **Safety** | Make sure students' hands are not too close to heat sources. When using a hair blower, tape down the cord so that it cannot be tripped over. Have students wear goggles. |

### What You Need:

☐ Crayons of 2 different colors
☐ Craft sticks
☐ Hair dryer
☐ Safety goggles

☐ Snack bags
☐ Aluminum cupcake tins
  (or metal measuring cups)

### How-to:

1. Peel the paper off your crayons. Now, these represent igneous rock—formed from cooled magma or lava.

2. Grate your "igneous rocks" with the craft stick to represent weathering. Sediment is formed by weathering.

3. Move your sediment and place the sediment in the snack bag. The process of carrying away sediment is erosion.

4. Press the sediment together to form sedimentary rock.

5. Put your sedimentary rock into the cupcake tin, and place it under a hair dryer to be changed by the heat. The rock that is formed from heat (and usually pressure) is metamorphic rock.

### Things to Think About:

1. What are the three types of rock? How do they form?

2. Sedimentary rocks are made from "parent" rocks. What is the parent rock for your sedimentary rock?

3. What do you think happens next to metamorphic rock?

**Instructions and General Information**

**Estimated Activity Time:** 10+ minutes

| Teacher Tips: OWL PELLETS | |
|---|---|
| Core Content | Ecosystems: Interactions, Energy, and Dynamics |
| Crosscutting Concepts | Stability and Change |
| Science and Engineering Practices | Analyzing and Interpreting Data |

| | |
|---|---|
| Guiding Question | How does energy flow through an ecosystem? |
| BEFORE the Event | Purchase sterilized owl pellets from a reputable vendor. You will need to find a sorting sheet online or with your purchased pellets to be used to identify the bones found in the pellet. Cover the tables with tablecloths. It is helpful to have this activity in a room with a sink or near bathrooms so participants can wash their hands after the activity. While we recommend Family Science Night events be free, this activity is very popular and can get quite costly. There are several ways to keep the cost down. First, you can recycle your pellets by collecting them back to reuse and let the fun be in the finding. Or you can also find a sponsor for this activity to cover the cost. Finally, you might want to sell tickets for a nominal fee (one to two dollars) to help with the expense and manage crowds. |
| DURING the Event | If recycling, put the extra bones back on the trays to be rediscovered. Participants can choose a few bones to tape onto their sorting sheets to keep. Make sure students wash their hands thoroughly after handling the pellets. |
| AFTER the Event | Make sure all the owl pellet leftovers are cleaned from floors, chairs, and tables. Wipe down all tables with antibacterial cloths. |
| Additional Resources | Have posters of different kinds of rodents that owls may eat. Include information about the food chain. |
| Things to Think About | 1. What items did you find most in the pellet? What was surprising? (Mostly bones. Encourage students to discuss what they found.)<br>2. What conclusions can you make about owls from the items you found? (The sorting sheet will provide information about what the owl eats.)<br>3. What does the owl's pellet tell you about how energy flows through an ecosystem? (Owls are predators and get their energy from other species.) |

| Activity: OWL PELLETS | |
| --- | --- |
| **Program Information Section**<br>*List location and hosts' names* | |
| **Science Behind the Activity** | Owls are consumers—they eat other animals for energy. Because they are at the top of the food chain, they are considered "tertiary" consumers. They are predators because they get their food from hunting. They cannot digest fur, bones, claws, beaks, and teeth from the animals they eat. These parts are formed into a tight pellet in their stomachs and spat out. | **FOOD CHAIN**<br><br>Energy Flows from Producers to Consumers<br><br>Owl (Tertiary Consumer)<br>↑<br>Rodent (Secondary Consumer)<br>↑<br>Insect (Primary Consumer)<br>↑<br>Plant (Producer) |
| **Vocabulary** | • Predators—organisms that live by hunting and eating others<br>• Prey—organisms that are hunted by predators<br>• Producers—organisms that make their own energy (plants are producers)<br>• Consumers—organisms that get their energy from eating producers and other (lower-level) consumers | |
| **Safety** | Have students wear eye protection (goggles) for this activity. We also recommend using sterile, latex-free gloves and washing hands with soap and water after the activity. Tell students to use caution when handling the dissection tools. Sanitize tabletop after completing this activity. | |

**What You Need:**

☐ Owl pellets

☐ Tweezers

☐ Toothpicks or meat skewers

☐ Trays

☐ Bone sorting chart (downloaded from online or supplied with pellets)

☐ Safety goggles

**How-to:**

1. Put on goggles and sterile gloves.

2. Dissect your owl pellet using tweezers and toothpicks.

3. Match bones and other items found in the pellet to the items on the sorting sheet.

4. Wash your hands when you are finished.

**Things to Think About:**

1. What items did you find most in the pellet? What was surprising?

2. What conclusions can you make about owls from the items you found?

3. What does the owl's pellet tell you about how energy flows through an ecosystem?

**Instructions and General Information**

**Estimated Activity Time:** 15 minutes

## Teacher Tips: PAPER AIRPLANES

| Core Content | Motion and Stability: Forces and Interactions |
|---|---|
| Crosscutting Concepts | Structure and Function |
| Science and Engineering Practices | Constructing Explanations and Designing Solutions |

| | |
|---|---|
| **Guiding Question** | How can you design a paper airplane to go the farthest distance? |
| **BEFORE the Event** | Set up a launching area that is away from the traffic flow. Mark the area with tape or cones. It makes it more fun to have a target at the end of the track. |
| **DURING the Event** | Make sure a designated host is monitoring the airplane launching. Sometimes the paper on the airplane is too thick to punch a hole. If so, tape a paper clip under the plane, hook a rubber band onto the paper clip, and launch. |
| **AFTER the Event** | Make sure all the furniture is back in place around the room and all paper airplanes are accounted for. |
| **Additional Resources** | Add other flying toys that demonstrate the forces of flight. You might invite someone to bring in a drone to demonstrate with this activity. Decorate the room as if it were an airport. |
| **Things to Think About** | 1. Which forces are at work to make your plane fly better? (Lift and thrust.)<br>2. Which forces are working against your plane? (Drag and weight [gravity].)<br>3. When was the force of thrust applied to your plane? (When it was launched.)<br>4. What changes did you make to your plane? Why did you make them? (Discuss with students how their designs might be improved to add lift or thrust, or to reduce drag [friction] or weight [the pull of gravity].)<br>5. How did the changes affect your airplane? (Answers may vary.) |

| | **Activity: PAPER AIRPLANES** |
|---|---|
| | **Program Information Section** |
| | *List location and hosts' names* |

| **Science Behind the Activity** | Airplanes fly on a level path when the forces of flight are balanced. Those forces are lift, weight (gravity), thrust, and drag. Lift and weight act in opposite directions, while drag and thrust oppose each other. A balance of these forces makes the plane move through the air smoothly. | **FORCES OF FLIGHT** *Lift* *Drag* *Flying Duck!* *Thrust* *Weight* |
|---|---|---|

| **Vocabulary** | • Force—a push or a pull<br>• Thrust—the forward force on a plane<br>• Drag—a force caused by friction of the air pushing on the plane, which holds back the plane and acts against thrust<br>• Lift—the air moving over and under the wings, which holds the plane in the air<br>• Weight—a measure of the force of gravity pulling the plane down |
|---|---|

| **Safety** | Students testing their airplane should wear goggles or some kind of eye protection. |
|---|---|

**What You Need:**

☐ Paper          ☐ Rubber bands          ☐ Scissors

☐ Scissors          ☐ Safety goggles          ☐ Hole punch

**How-to:**

1. Design and build your airplane. You can use a template or design your own.

2. Punch a hole at the front of the plane.

3. Put a rubber band through the hole.

4. Launch the plane from your finger.

5. Modify your design to improve it, based on your first try.

6. Launch your plane again.

**Things to Think About:**

1. Which forces are at work to make your plane fly better?

2. Which forces are working against your plane?

3. When was the force of thrust applied to your plane?

4. What changes did you make to your plane? Why did you make them?

5. How did the changes affect your airplane?

**Instructions and General Information**

**Estimated Activity Time:** 10 minutes

| Teacher Tips: RAINBOW STACKING | |
|---|---|
| Core Content | Structures and Properties of Matter |
| Crosscutting Concepts | Patterns |
| Science and Engineering Practices | Developing and Using Models |

| | |
|---|---|
| **Guiding Question** | Do all liquids have the same density? |
| **BEFORE the Event** | We use plastic specimen collection tubes that are purchased in bulk for this activity. But if you have connections with a medical lab, you might get them donated. Make sure to cover tables for easy cleanup. Have extra tablecloths in case of spills. Make sure all liquids are different colors. Before the event, use an oil-based candy dye to color the baby oil. We generally add green or blue dye to the baby oil and red food coloring to the water. Pour the liquids into cups on the table. Have the students hold their tubes over a paper plate in case they spill. Precut the pipettes to make a larger hole for the thicker liquids. |
| **DURING the Event** | Make sure the tubes are well sealed so they cannot be reopened. |
| **AFTER the Event** | Make sure all liquids are cleaned up from tables, chairs, and floors. |
| **Additional Resources** | Provide density sticks and blocks for students to explore the concept of density in solids. Have scales so students can weigh density sticks and cubes and see that even though the volume of objects can be the same, the mass can be different. |
| **Things to Think About** | 1. Which liquid is the least dense? How can you tell? (Baby oil; it floats to the top.)<br>2. Which liquid is the most dense? How can you tell? (Corn syrup; it sinks to the bottom.)<br>3. What will happen if you shake the liquids and mix them up? (They will settle back into their original layers.) |

## Activity: RAINBOW STACKING

### Program Information Section
*List location and hosts' names*

| Science Behind the Activity | Density is a very important concept in all areas of science. It describes how tightly the molecules in a substance are packed. More tightly compacted molecules have greater density than less-compacted molecules. Because density is the relationship of the mass of a substance to its volume, you can calculate density using this formula: <br><br> D = M/V (Density = Mass ÷ Volume) | Least Dense ↑ Most Dense |
|---|---|---|
| **Vocabulary** | • Density—how tightly or loosely the molecules are packed together | |
| **Safety** | Have students wear goggles and nonlatex aprons. Warn students not to drink liquids. Immediately wipe up any spilled liquid off the floor. Have students wash hands with soap and water after completing the activity. | |

**What You Need:**

☐ Water—colored with food coloring

☐ Corn syrup—clear

☐ Vegetable oil

☐ Baby oil—colored with green or blue oil-based candy dye

☐ Cups

☐ Pipettes

☐ Specimen tube with lid

☐ Tape

☐ Safety goggles

☐ Aprons

**How-to:**

1. Use a pipette to add the different liquids to your tube.

2. Put the lid on the tube.

3. Seal and tape the lid.

**Things to Think About:**

1. Which liquid is the least dense? How can you tell?

2. Which liquid is the most dense? How can you tell?

3. What will happen if you shake the liquids and mix them up?

**Instructions and General Information**

**Estimated Activity Time:** 10 minutes

| Teacher Tips: WORK IT, CIRCUITS! | |
| --- | --- |
| Core Content | Conservation of Energy and Energy Transfer |
| Crosscutting Concepts | Energy and Matter |
| Science and Engineering Practices | Planning and Carrying Out Investigations |

| | |
| --- | --- |
| **Guiding Question** | How does a circuit work? |
| **BEFORE the Event** | Old Christmas light strands work great for this activity, as well as for a student take-away. Ask parents for old unused light strings, and look at garage sales, thrift stores, and after-Christmas sales! Include some additional activities for students to explore at the station on their own, such as a basic circuits kit and energy balls. Discuss the need for eye protection. |
| **DURING the Event** | Use AA batteries. DO NOT directly connect the wires or clips on each side of the battery without the light bulb. Without the load (light bulb) the wire can heat up quickly and can burn fingers. |
| **AFTER the Event** | Leaving batteries connected runs down the battery quickly, so be sure to store your equipment with the batteries in a separate bag or storage bin. |
| **Additional Resources** | Provide an energy ball and circuits kit for students to explore further. |
| **Things to Think About** | 1. How does energy get from the battery to the light bulb? (It flows through the wire.)<br><br>2. How can you make the light work? (There are several correct ways to make a circuit. The wire must connect both the positive and negative terminals of the battery with the bulb along the path of electricity. The wire must touch both the bottom and side of the metal part of the bulb to work.)<br><br>3. What will happen if you put TWO or THREE light bulbs in the circuit? (This is a series circuit. Bulbs will be dimmer.)<br><br>4. What do you think you could do to make the light bulbs brighter? (One way to accomplish this is by putting another battery in the circuit. You can also guide students to build a parallel circuit with several pieces of wire. The bulbs will remain bright and not dim.) |

## Activity: WORK IT, CIRCUITS!

### Program Information Section
*List location and hosts' names*

| | | |
|---|---|---|
| **Science Behind the Activity** | Circuits are pathways that electrons flow through. A battery or other power source provides the electrons that move through the wires. As electrons move through the circuit, they can do work—for example, make a light bulb light up. For electricity to flow, the pathway must make a complete circuit and return to the source. Circuits can be constructed in series or parallel. Series circuits have a single path that flows through more than one device, while parallel circuits allow current to flow through multiple paths. | Simple Circuit |

| | |
|---|---|
| **Vocabulary** | • Current—the flow of electricity through a conductor<br>• Circuit—the path in which the electricity can follow; can be closed or open<br>• Closed circuit—has a complete path that allows the current to flow<br>• Open circuit—has a break in the path and the current cannot flow |
| **Safety** | Generally, there are no concerns using low-voltage batteries. Safety glasses or goggles are required for this lab. Use caution working with wires, as they can cut or puncture skin. Bulbs can shatter and cut skin. Be sure to tell students to NEVER experiment with electricity from wall outlets. |

**What You Need:**

☐ Light bulbs from Christmas lights with wire ends exposed
☐ Circuit ball
☐ Extra wire sections with the ends exposed
☐ Batteries (AA only)
☐ Safety goggles

**How-to:**

1. Make a complete circuit by using a light bulb and a battery.
2. Make a series circuit by connecting more than one light bulb in a row (twist the ends of the wires together).
3. Make a parallel circuit by twisting the wires from two light bulbs together with a third wire on one side and a fourth wire on the other. Then, attach the new wires to the battery.

**Things to Think About:**

1. How does energy get from the battery to the light bulb?
2. How can you make the light work?
3. What will happen if you put TWO or THREE light bulbs in the circuit?
4. What do you think you could do to make the light bulbs brighter?

**Instructions and General Information**

**Estimated Activity Time:** 10+ minutes

# Appendixes

In this section, we are sharing a variety of the forms, letters, and information that we have used over the years in our programs. This is not a complete record of our electronic resources but is designed to be a resource as you start planning your own program.

## Appendix A: Program Introduction

This introduction was inserted into our 2013 program, where middle school students presented 25-minute activities using a session model.

This year's event features more than 30 student-led sessions related to astronomy. From exploring how life would evolve on other worlds, to how we learn about newly discovered planets, each session will teach something about the universe beyond our world. Choose from any of these exciting sessions to discover more about space!

Dinner is served in the Celestial Café from 5:30 to 7:30. You can eat before your adventure begins at 5:30 or during one of the first three sessions. Be sure to visit the Book & Bake Sale to listen to some songs written by our students and have a beverage with dessert while you browse our Book Fair. We've also set up a "Science Surplus Store" this year and are selling raffle tickets for exciting prizes!

In each session you'll make something fun, eat something tasty or take home something interesting related to each session. Sessions start every half hour and last for 25 minutes. There is a five-minute transition time between sessions to get to your next stop. You are guaranteed to have fun, no matter which sessions you choose! Check the schedule to see times and locations!

## Appendix B: Letters to Middle School Hosts' Parents

This letter and order form were sent home only to parents of the students who were involved in hosting the middle school program using a session model. They provided information about student responsibilities and the event.

---

**Family Science Night 2012: Ocean Adventure!**

Parents,

Six years ago, we started our first Family Science Night event, which has been an annual occurrence ever since. This year, "Ocean Adventure" is the theme for the program your students will host. This exciting event has brought in hundreds of people from the school and community each year to engage in science-related activities. Our students showcase their talents while sponsoring the event. This is one of the highlights of the year for our students, who always seem to enjoy the event while gaining leadership experience. It also gives them a chance to give back to the school and community.

Here's what you need to know about this year's event:
- **Date:** Monday, March 12
- **Student time:** Students who are participating will stay on campus from dismissal until 9:30 p.m.
- **Event time:** The event will start with a 5:30 p.m. dinner session; student sessions will run every half hour starting at 6:00 p.m. and ending at 9:00 p.m.

Students will have a range of responsibilities and assignments for the evening. They include the following:
- Session presenters, managers, photographers, and more
- Hosts and welcome station for our guests
- Setup and cleanup

All students will need to plan on helping with decorations, either in their assigned room or general areas. We ask for students and parents to help with making, securing, or buying decorations for each assigned area. Please sign and return this letter indicating permission and availability for your child to participate in sponsoring this event. On the back is additional information about dinner, T-shirts, and other opportunities. As always, please let one of us know if you have any questions. Thank you!

❏ I give permission for my child to participate in hosting this year's Family Science Night on March 12.

❏ My child will NOT be able to participate in hosting this year's Family Science Night.


_____  _____  _____
Student's Name              Parent Signature              Date

---

## Ocean Adventure: Additional Information

| Dinner Time | T-shirts |
|---|---|
| We will provide student meals for $5.00. Students will eat between 5 and 5:30. Please send either $5.50 for your student's meal or a sack dinner. You may preorder meals for other family members. Meals are $5.50 in advance, $6.50 @ door. | Students are asked to purchase a T-shirt for the event. These are sold to students at close-to-cost for $10.00. Student T-shirts will have "STAFF" printed on the back. Additional T-shirts can be purchased for other family members for $15. |
| **Portable Planetarium** | **Evening Highlights** |
| The portable planetarium is returning this year! While your child is probably going to be too busy to go to this session, you or other family members may want to. Tickets are $2 in advance, $3 at the door. | Coffee Shop<br>Bake Sale and Book Fair<br>Over 25 student-led activity sessions! |
| **Adopt-a-Pirate** | **Program Advertising** |
| Our mascots are again being offered for adoption. These 5-foot, inflatable pirates are used to hold signs at each session location. The adoptions cost $15.00. Students get to name their pirate, get a certificate of adoption and will take their pirate home at the end of the evening. | This year we will again have an expanded program guide including a map of the school, more information about each session and sponsor advertising. Advertising costs will be $30 for 1/2 page. Double spaces, full page, are available for $50. Ad copy *MUST* be received by March 1 to be included! |

### Family Science Night Payment – Please submit NO LATER THAN 3/1

❑ $5.50 for Burgers Dinner
  Check one: ❑ hamburger ❑ hot dog -OR- ❑ Will Bring Dinner
❑$10 for student T-shirt
  Check one: ❑Adult Small ❑Adult Med ❑Adult LG ❑Adult XL
**Additional payment amount included for:**

❑$15 for each additional T-shirt – Total number: _____

Check size(s): ❑Adult Small ❑Adult Med ❑Adult LG ❑Adult LG ❑Adult XL ❑Adult 2X

❑$2 for each planetarium ticket paid for in advance: Total number: _____ X $2 = _____

❑ $5.50 for Additional Meals: Number of each: _____hamburgers _____hot dogs
❑$15 for pirate adoption – to be named: _____

❑$30/$50 each sponsor advertisement –Total number: _____ - SEND AD COPY by 3/1
        **TOTAL AMOUNT ENCLOSED:** _____

### NOTE: Parent volunteers will also be needed for the Coffee Shop and Bake Sale:

❑ I'm available! Name: _____ Email: _____

## Appendix C: Student Session Planning Form

Middle school students who participated as activity hosts in events were required to complete a planning form for each session. Groups met with the teacher early in the planning process for advice and recommendations. The middle school program always used a session model.

**Night of Discovery!**

**Family Science Night Presentation Summary**

Session Title: _____

Name of Discover: _____

Discovery: _____

Group Members:

_____          _____

_____          _____

Mascot number we are responsible for: _____

Abstract (Summary of Presentation): _____

_____

_____

_____

Introduction Format (5 minute overview):

____ Powerpoint     ____ Movie Maker     ____ Other: _____

Description of our introduction: _____

_____

_____

Description of our demonstration: _____

_____

Description of our activity: _____

_____

_____

_____

**Thinking Ahead**

Decorating Plans: _____

_____

What materials do we need to get to decorate with? _____

_____

What will your participants leave with that makes your session special?

___ Trinket       ___ Make & Take       ___ Food

Description of food/trinket/make-and-take and why it is appropriate for your session:

_____

_____

What supplies do we need to conduct our session? _____

_____

_____

What items does our GROUP need to obtain? _____

_____

_____

What will need to have copies made for? _____

What items do we need for the TEACHERS to obtain for us? _____

_____

_____

What responsibilities does each team member need to take charge of?

_____   _____

_____   _____

_____   _____

_____   _____

# Appendixes

## Appendix D: High School Science Ambassador Application

High school students who wanted to participate in the club were asked to complete an application. This process ensured parent consent and a commitment to the program.

### SCIENCE AMBASSADOR APPLICATION

**Who:** Students passionate about science, teaching and/or working with elementary students.

**What:** Volunteers needed to run Science Night at neighboring elementary schools.
We need our high school students to lead science activities at these events! A few examples of the activities to be offered include: film canister rockets, Cartesian divers, rain makers, singing straws, color crystals, roller coasters and more!
Come help students navigate through the science and engineering design process!

**Cost:** $35.00 per participant cost include T-shirts AND pizza dinner on all 6 events.

**Duties:** Help run activities, greet families at the door, hand out bags, and other activities to help the event run smoothly.

**When:** Informational Meeting - (3:45–4:15) on August 25;
First regular club meeting September 22 (3:45–4:30)
Monthly after school meetings will be held on the last Tuesday of each month through December to prepare for the elementary school events. Elementary school events will be held on Tuesdays from January 12 through March. You MUST be available every Tuesday during that time period unless prior arrangements are made.

**Why:** Share your love of science, gain community service hours, build your college resume and meet like-minded individuals in your school and community.

---

**How do I apply?** Complete the form below, ask a science teacher for a recommendation (back).

**Due Date:**          **September 15**          You will be notified of acceptance by email by Sept. 18.

**Step 1: Complete form below:**

NAME: _____     EMAIL:_____

I confirm I am currently passing my science class and I have not earned a discipline referral this

year. I will be available every Tuesday for events January, February and March.


_____Date: _____ _____Date: _____
Student signature                                              Parent Signature


**Step 2:** Science teacher recommendation – *Returning members do NOT need this.*
**Step 3:** Follow-up with teacher to ensure your recommendation has been turned in.
**Step 4:** Plan to stay for the first 9/29 meeting!

**Teacher Recommendation for Science Ambassadors**

ON A SCALE OF 1 TO 5 (5 being the highest rating) rate the candidate on the following items

Responsibility: _____

Works well in a group: _____

Maturity level: _____

Self-motivated: _____

Would you recommend this student to work with elementary age students? YES or NO

Do you have any reservations about recommending this candidate to work a booth with young children? Please comment below:

_____

_____

_____

_____

Do you have any additional comments or suggestions about this student? Please comment below:

_____

_____

_____

_____

Teacher name: _____(print)

Teacher signature:

_____Date:_____

THANK YOU FOR YOUR TIME.

PLEASE PLACE THE FORM IN Dr. Governor's box or deliver in person.

## Appendix E: Sponsor Request

Letters went out to potential sponsors for advertising in the program early in the planning process each year.

### High School Science Ambassadors
#### *Sponsor Opportunities*

Our **Science Ambassadors** program is partnering with local elementary schools to put on a Family Science and Engineering Night for elementary students and their families. This night brings students and their families to the school to experience the fun of science and engineering. Our mission is for all students prek-5th grades and their families to engage in science and engineering activities that have real world connections to develop critical thinking and problem-solving skills. Our goal is to inspire more students with the confidence to pursue careers in the STEM (Science Technology Engineering Math) fields.

A full book program with a list of all activities will be distributed at **all six** elementary school events run by the Science Ambassadors for the 2015-16 school year. Program size is 4.25" x 5.5".

During the 2014-15 school year our event reached thousands of families at **all six** community elementary schools! Don't miss out on this opportunity to promote your business in the local community!

Advertising opportunities will be available as follows:

Whole Page advertisement . . . . . . . . . . . . . . . . . $250     Full page behind event pages
Half Page advertisement . . . . . . . . . . . . . . . . . . . $150     ½ page behind event pages
Activity sponsor . . . . . . . . . . . . . . . . . . . . . . . . . . $ 75     Business card sized space

Any combination of $300 or more in advertisements will include your (business) name on the event banner.

**This Year's Schedule:**
    January 12, January 26, February 9, March 8, March 15, March 22

**Contact information:**
- Dr. Donna Governor    (email)
- Ms. Denise Webb      (email)

Return your check and advertising copy to Donna Governor no later than _____..
Make checks payable to the High School, please.

## Appendix F: Event Flyer 1

These event flyers were sent home to parents of all students at the middle school where Family Science Night events were held (using a session model).

---

**ANNOUNCING our 2013 Family Science Event:**
**Night of Discovery**
**March 18**
**5:30–9:00 p.m.**

Plan now for an evening of fun and adventure with
famous discoverers from the History of Science!
Come join us as our students present an evening of science fun!
RESERVATIONS **NOT** REQUIRED
ALL EVENTS FREE! (Except planetarium & dissection)
<u>This year's program includes:</u>
- Forty student-led, activity-rich sessions to choose from beginning at 6:00
  - Sessions repeat every half hour

**Session Sampler:**
- Wright Brothers - Flight
- Walt Disney - Animation
- Nikola Tesla - Electricity
- Marie Curie - Radiation
- Watson & Crick - DNA
- Neil Armstrong – Lunar Surface
- Werner von Braun - Rockets
- Robert Ballard - Titanic
- Discovery Café (Bake Sale & Book Fair)
- ........And More!

**Other Activities:**
- Dinner from Backyard Burgers beginning at 5:30
- $5.50 in advance, $7.00 at the door – Burgers or Hot Dogs
- Ticketed Events – LIMITED SEATING:
  - Portable Planetarium Shows - $2 in advance, $3 at the door
  - Owl Pellet Dissection - $2 in advance, $3 at the door
- Coffee shop with Bake Sale & Book Fair
- Discovery Event T-shirts - $15 (advance orders only)
- Opportunity for your business to advertise in our event program
  email Dr. Donna Governor for more information

## Night of Discovery: Event Preorder Form

| Dinner Time | T-Shirts |
|---|---|
| Dinner will be available from 5:30 to 7:30. Menu: Hot Dogs & Burgers Meals: $5.50 in advance, $7.00 @ door. | These special blue event T-shirts are $15.00 and are available in adult or youth sizes. No shirt orders after 3/1! |
| **Portable Planetarium** | **Program Advertising** |
| The portable planetarium is returning this year with a special show on astronomical discoveries! | This year we will again have an expanded program guide. Advertising costs are $30 for 1/2 page, full page for $50. |

### EVENT PRE-ORDERING FORM

❑ $5.50 for Prepaid Dinners - Total Number: _____

    Indicate number of each: ____ Hamburgers ____ Cheeseburgers ____ Hot Dogs

❑ $15 for T-shirts - Check one for each shirt ordered – Total Number: _____

    ❑Youth Small ❑Youth Med ❑Youth LG ❑Youth XL

    ❑Adult Small ❑Adult Med ❑Adult LG ❑Adult XL ❑Adult 2X

Planetarium & Owl Pellet Dissections are $2.00 in advance, $3.00 at the door

For each indicate time preferences for first, second & third choice session times:

❑ Planetarium: Total number: _____ X $2 = _____

    ___ 6:00 ___ 6:30 ___ 7:00 ___ 7:30 ___ 8:00 ___ 8:30

❑ Owl Pellet Dissection – PARENT MUST ACCOMPANY - Total: ___ X $2 = _____

    ___ 6:00 ___ 6:30 ___ 7:00 ___ 7:30 ___ 8:00

    TOTAL AMOUNT ENCLOSED: _____

**Make checks payable to the school & return to Dr. Donna Governor**

**ALL PRE-PAID ITEMS WILL HAVE TICKETS/SHIRTS READY AT THE DOOR FOR PICK UP THE NIGHT OF THE EVENT**

EMAIL DR. GOVDERNOR FOR MORE INFORMATION OR IF CONFIRMATION DESIRED FOR PREORDERS AFTER PAYMENT SENT IN

*ALL PREPAID ORDERS MUST BE RECEIVED BY MARCH 1 FOR LOWER PRICES*

*FULL PRICE MEALS, DISSECTION & PLANETARIUM TICKETS AVAILABLE @ DOOR*

## Appendix G: Event Flyer 2

These flyers were sent home to elementary school students to advertise events at their schools.

---

**Family Science & Engineering Night**
Tuesday February 24
At Our Elementary School

**PTSO Meeting:**

6:00 p.m. in the Cafeteria

**Activity Stations Open:**

6:30 p.m.

**FREE Hands-on Fun for the Whole Family!**

**\*Experiments:** Lava Lamps, Static Sensations, Plant Dissections, Work-it-Circuits, Chemical Reactions & MORE!

**\*Engineering Challenges:** Hall Rollers, Roller Coaster Challenge, Foil Boats & MORE!

---

# Appendixes

## Appendix H: Checklists

We've copied and pasted all our checklists here into the appendix for easy reference.

# Checklist 2.1.

## Planning Timeline

Based on our experiences, here's our suggested timeline for planning. If you plan to hold your event in the fall, you will need to accelerate the timeline.

**September**
- ☐ Get administrative approval
- ☐ Set a date
- ☐ Decide event format
- ☐ Decide who will run the activities
- ☐ Recruit activity hosts (students, volunteers, or teachers)

**October**
- ☐ Plan events and activities
- ☐ Determine spaces to be used
- ☐ Recruit adult monitors
- ☐ Compile supply list
- ☐ Invite related groups to participate
- ☐ Arrange ticketed activities and sessions
- ☐ Hold fundraisers
- ☐ Find advertisers for program
- ☐ Arrange bus transportation for hosts if using older students for events in elementary schools

**November**
- ☐ Order supplies
- ☐ Schedule practice sessions for activity hosts
- ☐ Review any student-developed materials
- ☐ Assign additional roles (e.g., photographers)
- ☐ Prepare a draft program

**December**
- ☐ Organize and prepare materials
- ☐ Rehearse presentations
- ☐ Finalize program

**January**
- ☐ Print programs
- ☐ Hold final activity practice

# Checklist 3.1.

## Student Volunteer Requirements

When working with students, make sure that they complete the following activities:

- ☐ Attend an orientation meeting
- ☐ Obtain parental permission to participate
- ☐ Agree to participation requirements
- ☐ Commit to event date
- ☐ Prepare activity or presentation
- ☐ Attend work sessions to prepare
- ☐ Submit a supply list of required supplies
- ☐ Practice and rehearse the presentation or activity
- ☐ Make arrangements for transportation home
- ☐ Decorate and set up space as necessary for activity or presentation
- ☐ Clean up activity space after event

# Checklist 3.2.

## Activity Host Responsibilities

There is some overlap on this list with the checklist for student volunteer responsibilities; however, this list is more specific to the activity host role.

☐ Determine activity
☐ Review content for mastery
☐ Write abstract for program
☐ Prepare presentation, if applicable
☐ Practice presentation or activity
☐ Request necessary materials
☐ Organize materials and bring to event
☐ Determine and arrange decorations for your session
☐ Set up for the event
☐ Create a fun atmosphere for your session
☐ Interact with children at event
☐ Clean up
☐ Request additional materials, if applicable, for future events
☐ Store or return materials

# Checklist 3.3.

## Monitor Responsibilities

If assigned to an activity room:

☐ Introduce yourself to the activity hosts
☐ Monitor attendance to avoid activity exceeding capacity
☐ Handle any disruptions from parents and children attending the activity
☐ Report any problems with activity hosts to the event coordinator
☐ Dismiss activity hosts at the end of the event after the space has been cleaned up

If assigned to a common area (e.g., hall, cafeteria):

☐ Monitor attendees moving from activity to activity
☐ Handle any disruptions from parents and children attending the activity
☐ Report any problems to the event coordinator

# Checklist 3.4.

## Greeters

☐ Decorate the entrance to set the mood for the event
☐ Become familiar with the event so that you can answer questions
☐ Distribute programs and bags to attendees—usually one per family
☐ Facilitate pickup or selling of tickets for special activities
☐ Greet attendees with a smile
☐ Clean up the entrance at the end of the event

# Appendixes

## Checklist 3.5.

### Chaperones

- ☐ Decorate the space the guest will use to set the mood for the event
- ☐ Welcome the guest upon arrival
- ☐ Offer the guest water or refreshments
- ☐ Help the guest set up
- ☐ Monitor and help with traffic in and out of the guest's space
- ☐ Clean up the entrance at the end of the event

## Checklist 3.6.

### Photographers and Videographers

- ☐ Take photos/videos of all preparation activities
- ☐ Obtain parent permission to photograph students if necessary
- ☐ Confirm that your equipment is fully charged prior to the event
- ☐ Make sure you have backup batteries on hand
- ☐ Be sure to get images/video of all activities
- ☐ Focus on participant engagement in your photos/video
- ☐ If taking photos, review and edit photos for composition
- ☐ If producing a video, edit and add credits
- ☐ Submit finished product/portfolio to coordinator within one week

## Checklist 3.7.

### Managers

- ☐ Identify activity groups that require assistance for the event sponsor
- ☐ Assist in ordering supplies
- ☐ Inventory and organize supplies as they arrive
- ☐ Track volunteer hours of all participants
- ☐ Assist in preparing event programs
- ☐ Oversee and distribute common materials (e.g., table covering, tape, brooms).
- ☐ Circulate at the event and assist in supply issues
- ☐ Advise event coordinator of any issues that need to be addressed
- ☐ Verify that all spaces used during the event have been cleaned and returned to the original (or better) condition
- ☐ Prefrom other tasks as assigned

National Science Teachers Association

# Checklist 8.1.

## Hosting School's Responsibilities

- ☐ Obtain permission from the administration
- ☐ Set a date that will work for both your school and the visiting hosts
- ☐ Arrange for extra personnel such as school resource officers and custodians
- ☐ Advertise your event to your school
- ☐ Coordinate with the program manager to assign spaces and create a map for the program
- ☐ Conduct fundraising to contribute to the cost of the program
- ☐ Plan for extra activities to be included (e.g., meals, bake sale, career fair)
- ☐ Arrange for a reception area with greeters
- ☐ Make sure your attendees understand that the hosts are guests and that the attendees should be patient with them
- ☐ Provide radios for communication
- ☐ Make available necessary equipment such as ice machines, refrigerators, and cleaning supplies
- ☐ Help clear the building at the end of the event
- ☐ Send thank-you notes to your guest hosts

# Checklist 8.2.

## Organizer's Responsibilities

- ☐ Set dates with contacts at other sites (plan weather backup dates as well)
- ☐ Plan the core activities and sessions
- ☐ Recruit hosts and obtain parental permission if working with students
- ☐ Assign sessions to host groups
- ☐ Provide planning and practice sessions for your hosts
- ☐ Coordinate with other school programs to make sure that students will not have any conflicts on the event nights
- ☐ Work with your contact to arrange activity spaces; communicate any special requirements for specific activities
- ☐ Assemble and print programs that can be used at multiple locations
- ☐ Arrange for shirts and name tags for hosts
- ☐ Feed your hosts prior to the event
- ☐ Transport your hosts to the site
- ☐ Gather, store, and transport materials and supplies—replenish as necessary
- ☐ Make sure that your hosts clean up at the end of the event and are picked up by their parents
- ☐ Arrange for e-mail and group communication (possibly using a group app) for sharing information

# Appendixes

## Appendix I: Activity Index

| Activity | Level | Core Content | Crosscutting Concepts | Science and Engineering Practices |
|---|---|---|---|---|
| Air Cannon | Intermediate | Conservation of Energy and Energy Transfer | Energy and Matter | Planning and Carrying Out Investigations |
| Balancing Bugs | Novice | Conservation of Energy and Energy Transfer | Structure and Function | Developing and Using Models |
| Balloon-Powered Cars | Advanced | Forces and Motion | Cause and Effect | Constructing Explanations and Designing Solutions |
| "Bear-y" Hot S'mores | Intermediate | Conservation of Energy and Energy Transfer | Energy and Matter | Analyzing and Interpreting Data |
| Bubble Olympics | Novice | Structure and Properties of Matter | Structure and Function | Planning and Carrying Out Investigations |
| Bug Buzzers | Intermediate | Sound Waves | Cause and Effect | Planning and Carrying Out Investigations |
| Cartesian Divers | Advanced | Motion and Stability: Forces and Interactions | Cause and Effect | Developing and Using Models |
| Catapults | Novice | Conservation of Energy and Energy Transfer | Cause and Effect | Constructing Explanations and Designing Solutions |
| Catch the Wave | Novice | Earth Materials and Systems | Energy and Matter | Developing and Using Models |

*Continued*

National Science Teachers Association

Appendix I *(continued)*

| Activity | Level | Core Content | Crosscutting Concepts | Science and Engineering Practices |
|---|---|---|---|---|
| **Changing-Color Slime** | Novice | Structure and Properties of Matter | Stability and Change | Constructing Explanations and Designing Solutions |
| **Chemical Reactions** | Advanced | Matter and Its Interactions | Energy and Matter | Planning and Carrying Out Investigations |
| **Dino Discovery** | Intermediate | Evidence of Common Ancestry and Diversity | Stability and Change | Engaging in Argument From Evidence |
| **Face Magnets** | Intermediate | Forces and Interactions | Cause and Effect | Planning and Carrying Out Investigations |
| **Foil Boats** | Novice | Conservation of Energy and Energy Transfer | Structure and Function | Constructing Explanations and Designing Solutions |
| **Getting Buggy** | Novice | Natural Selection and Adaptation | Structure and Function | Constructing Explanations and Designing Solutions |
| **Grassy Pets** | Novice | Cycles of Matter and Energy Transfer in Ecosystems | Structure and Function | Planning and Carrying Out Investigations |
| **Hall Rollers** | Advanced | Conservation of Energy and Energy Transfer | Energy and Matter | Planning and Carrying Out Investigations |

*Continued*

Appendix I (continued)

| Activity | Level | Core Content | Crosscutting Concepts | Science and Engineering Practices |
|---|---|---|---|---|
| Harmony Harmonicas | Novice | Waves: Light and Sound | Cause and Effect | Planning and Carrying Out Investigations |
| Hovercraft | Advanced | Forces and Motion | Cause and Effect | Developing and Using Models |
| Ice Cream | Novice | Conservation of Energy and Energy Transfer | Energy and Matter | Planning and Carrying Out Investigations |
| Krazy Kaleidoscopes | Advanced | Light | Structure and Function | Constructing Explanations and Designing Solutions |
| Lava Lamps | Intermediate | Structure and Properties of Matter | Cause and Effect | Developing and Using Models |
| Maracas | Novice | Conservation of Energy and Energy Transfer | Cause and Effect | Planning and Carrying Out Investigations |
| Mighty Lungs | Advanced | Structure and Processes | Structure and Function | Developing and Using Models |
| Modeling the Rock Cycle | Advanced | Earth's Materials and Systems | Stability and Change | Developing and Using Models |
| Owl Pellets | Advanced | Ecosystems: Interactions, Energy, and Dynamics | Stability and Change | Analyzing and Interpreting Data |

Continued

National Science Teachers Association

Appendix I (*continued*)

| Activity | Level | Core Content | Crosscutting Concepts | Science and Engineering Practices |
|---|---|---|---|---|
| Paper Airplanes | Advanced | Motion and Stability: Forces and Interactions | Structure and Function | Constructing Explanations and Designing Solutions |
| Parachutes | Intermediate | Force and Motion | Structure and Function | Constructing Explanations and Designing Solutions |
| Pinwheels | Novice | Earth's Systems | Scale, Proportion, and Quantity | Developing and Using Models |
| Plant Dissection | Intermediate | Structure and Processes | Structure and Function | Analyzing and Interpreting Data |
| Rainbow Stacking | Advanced | Structure and Properties of Matter | | Developing and Using Models |
| Roller Coaster Challenge | Intermediate | Force and Motion | Patterns | Developing and Using Models |
| Static Sensations | Intermediate | Conservation of Energy and Energy Transfer | Energy and Matter | Planning and Carrying Out Investigations |
| Straw Rockets | Intermediate | Force and Motion | Energy and Matter | Constructing Explanations and Designing Solutions |
| Surface Tension | Intermediate | Structure and Properties of Matter | Structure and Function | Planning and Carrying Out Investigations |
| Work It, Circuits! | Advanced | Conservation of Energy and Energy Transfer | Energy and Matter | Planning and Carrying Out Investigations |

# Index

Page numbers printed in **bold** type indicate tables, figures, or illustrations.

# Index

# Index

dissection activities, 21, 45, 47
Dodd, Robert, 44
donations, 67
donor appreciation, **79,** 79–80

## E

Earth's Materials and Systems
    Catch the Wave activity, 102–103, 188
    Modeling the Rock Cycle activity, 162–163, 190
Earth's Systems
    Pinwheels activity, 118–119, 191
Ecosystems: Interactions, Energy, and Dynamics
    Owl Pellets activity, 164–165, 190
Emmitt, Chris, 8
Energy and Matter concept
    Air Cannon activity, 122–123, 188
    "Bear-y" Hot S'mores activity, 124–125, 188
    Catch the Wave activity, 102–103, 188
    Chemical Reactions activity, 152–153, 189
    Hall Rollers activity, 21, **88,** 154–155, 189
    Ice Cream activity, 114–115, 190
    Roller Coaster Challenge activity, 138–139, 191
    Static Sensations activity, 140–141, 191
    Surface Tension activity, 144–145, 191
    Work It, Circuits! activity, 90, 91, 170–171, 191
Engaging in Argument From Evidence
    Dino Discovery activity, 128–129, 189
event-day details
    during event, 73–75
    last-minute items, 69–72, **71**
    post-event cleanup, 75–76
Evidence of Common Ancestry and Diversity
    Dino Discovery activity, 128–129, 189

## F

Face Magnets activity, 130–131, 189
Family Science Nights
    approaching administration, 19–20
    choosing a date, 20
    event day details, 69–76
    event programs, 48–50, 48f, 49f, 50f

# Index

# Index

# Index

# Index

**W**